PART P – ELECTRICAL DOMESTIC INSTALLATIONS FOR DEFINED SCOPE OPERATIVES

Including England and Wales

STUDY NOTES

E40

Published by ConstructionSkills, Bircham Newton, King's Lynn, Norfolk, PE31 6RH

© **Construction Industry Training Board 2005**

The Construction Industry Training Board otherwise known as CITB-ConstructionSkills and ConstructionSkills is a registered charity (Charity Number: 264289)

First published August 2005

ISBN: 978-1-85751-176-5

A joint publication between

National Association of Professional Inspectors & Testers

CITB-ConstructionSkills is a partner in ConstructionSkills and from now on you will see more and more of our products and services branded ConstructionSkills. To find out why and what this means to you please visit www.cskills.org

ConstructionSkills has made every effort to ensure that the information contained within this publication is accurate. Its content should be used as guidance material and not as a replacement for current regulations or existing standards.

All rights reserved. No part of this publication may be reproduced, stored in a retrieval system or transmitted in any form or by any means, electronic, mechanical, photocopying, recording or otherwise, without the prior permission in writing from ConstructionSkills.

Printed in the UK

For a comprehensive listing of all BES publications turn to the back page.
Tel: 01485 577800   Fax: 01485 577758   E-mail: publications@cskills.org

ACKNOWLEDGEMENTS

ConstructionSkills wishes to express its thanks to the following for their assistance in the productions of this manual.

Consultant

NAPIT (K Lorriman)

Others

Elm Training Services

Techtrain Associates

Polar Pumps Training

Note: This manual contains abbreviated extracts and paraphrases of the IEE regulations. It is emphasised that these interpretations of the regulations have been devised for the purpose of training and should not be regarded as authoritative in any other context. When necessary, the regulations should be referred to directly.

CONTENTS

	Page no.
Introduction	1
Electrical Concepts	5
Regulations and Standards	25
Plan and Style of Regulations BS 7671	33
Scope, Object and Fundamental Principles (*Part 1*)	36
The Building Regulations 2000 for England and Wales	44
Lighting, Power Circuits and Heating System Controls	45
Lighting Circuits	50
Power Circuits	57
Heating System Controls	67
Equipment and Accessories	73
Residual Current Devices (RCDs)	87
Earthing Arrangements and Protective Conductors	90
PVC Wiring, Installation Techniques and Cable Selection	107
Harmonised Cable Core Colours	111
Cable Selection	128
Armoured Cables	131
Installation Practices to BS 7671	138
Safe Isolation	166
Safe Isolation Procedure	173
Inspection and Testing	174
Appendices	203 |//
Appendix A NAPIT - Loop Impedance Test Values	
Appendix B NAPIT - Minor Works Certificate	
NAPIT Initial Application Form	

INTRODUCTION

Section 1 of Approved Document P of the Building Regulations 2000 (England and Wales) covering electrical safety, often referred to as Part P, clearly states that electrical installations should be:

- designed
- installed
- suitably enclosed

In order to provide protection for persons against the risks of electric shock, burns or fire injuries.

In order to achieve this, electrical installations should be inspected and tested during and at the end of the installation, before being used, to verify that they are reasonably safe. This is usually achieved by ensuring that installation complies with BS 7671 (the IEE Wiring Regulations).

A way of demonstrating this compliance would be to follow the procedures given in Chapter 74 of BS 7671: 2001.

- Supply to the person ordering the electrical installation work copies of the appropriate inspection and test certificate, signed by a person competent to do so.

- In the case of a competent person, registered with an electrical self-certification scheme, to the building control body, a declaration that compliance with the building regulation has been achieved.

The Test certificate should show that the electrical installation work has been:

- inspected (during and on completion) to verify that the components (cables and accessories) are:
  - made in compliance to British Standards or European Standards
  - selected and installed in accordance with BS 7671
  - not visibly damaged or defective

- tested to check satisfactory performance in relation to:
  - continuity of conductors
  - insulation resistance
  - separating of circuits
  - polarity
  - earthing and bonding arrangements
  - earth fault loop impedance
  - functionality of all protective devices including residual current devices

For defined scope electrical installation work the following types of certificates should be used:

Level B   The installation of a new circuit from consumer unit.

        An electrical installation certificate should be completed and issued.

The following pages contain copies of these certificates. In order to assist in understanding what is required to complete them they have been marked to indicate where the information may be obtained from in this publication.

At the end of this publication are examples of completed certificates.

Note: Where sections of the building regulations affect installation methods the following identification is used to indicate the part of the building regulations e.g. $\boxed{M}$ *indicating access to and use of building.*

Minor Works Electrical Certificate

For compliance with Building Regulations Part P

NOTES:

1. This Minor Works Electrical Certificate **shall only be used** for the reporting on the condition of an electrical installation, where the work does not comprise the addition of a new circuit.
2. This Certificate is based upon the format of the Minor Electrical Installation Works Certificate, issued by the Institute of Electrical Engineers and published in BS7671:2001
3. **The Inspection, Test, Verification of the installation and Completion of this certificate must be undertaken by a person on a Competent Persons Scheme approved by the Secretary of State or on the NAPIT Register of Approved Electrical Inspectors.**

NAPIT Minor Works Electrical Certificate.

Information for recipients (to be appended to the report).

This Minor Works Electrical Certificate **shall only be used** for the reporting on the condition of an electrical installation, where the work does not comprise the addition of a new circuit.

You should have received an original Report and the contractor should have retained a duplicate.

If you were the person ordering this Report, but not the owner of the installation, you should pass this Report, or a copy of it, immediately to the owner.

The original Report is **to be retained in a safe place and be shown to any person inspecting or undertaking work on the electrical installation in the future.**

If you later vacate the property, this Report will provide the new owner with details of the condition of the minor electrical installation works at the time the Report was issued.

For safety reasons, the electrical installation will need to be re-inspected at appropriate intervals by a competent person.

The maximum time interval recommended before an inspection of the installation is stated in the Report under 'An Inspection' [5(4)] and the Institute of Electrical Engineers recommend every 10 years or on change of occupancy for Domestic Electrical Installations

If this work is carried out in a dwelling, you should also receive a 'Compliance with Building Regulations Declaration' **within 30 days of the electrical installation being completed.**

Caution:

An initial assessment of the existing electrical installation must be made to ensure that the proposed Minor Electrical Installation Works can be undertaken safely, this assessment must be undertaken by a person competent to do so.

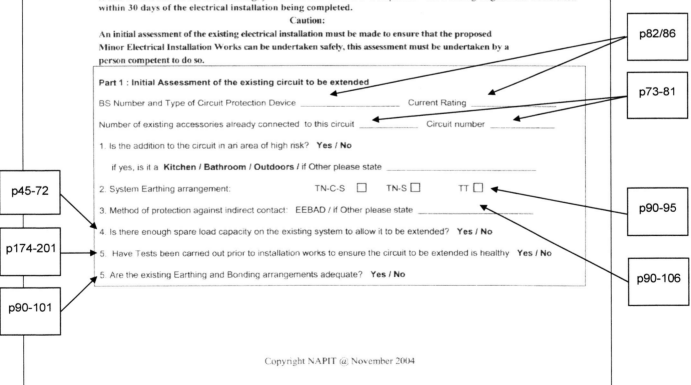

Copyright NAPIT @ November 2004

Minor Works Electrical Certificate

NAPIT Minor Works Electrical Certificate (note 1)

Only to be used for minor electrical work, this does not include the provision of a new circuit.

Part 2 : Description of the minor works

1. Description of the minor works : _____ Located _____
2. Address of the property: _____

4. Date the minor works completed : _____

→ p45-72

→ P97-106

Part 3 : Visual Check at 'First Fix' Stage (1st Column) Inspection Results (2nd Column) (tick, cross, N/C or N/A)

Main Equipotential Bonding present and adequate		All Cable cores correctly identified in joints and accessories	
Correct Circuit Protection Device fitted and identified		Appropriate Supplementary Bonding present and adequate	
Correct Cable type and size used, allowing for external influences		Modifications to the Building Fabric appropiate and safe	
Cable run in 'safe' zones or mechanically protected		All Accessaries correctly placed as per Approved Document M and BS 8300	
Cables securely fastened or in appropriate wiring protection systems		All covers replaced, Accessories secure and neatly aligned.	
All cable joints correctly terminated, secure and accessable		Circuit details updated on the installation schedule (or schedule produced)	
Name	Signed	Position	Date

← p82-89

← p57, p107-113, p128-130

← p144-165

← p122-127

→ p106-113, p139-140

→ p148-161

→ p76-78

→ p141-142

Part 4 : Essential Tests

1. Earth Continuity: Value (R1+R2 or R2) _____ Ω, If R2 is this Satisfactory? **Yes / No** ← p178-180

2. Insulation resistance:

   Phase/Neutral .... MΩ

   Phase/Earth .... MΩ ← p184-186

   Neutral/Earth.................. MΩ

3. Polarity : Satisfactory **Yes / No** ← p186-188

4. Earth fault loop impedance Ω

5. RCD (if applicable) : Rated residual operating current $I_{\Delta n}$ ....mA and operating time of ....ms (at $I_{\Delta n}$) ....ms (at $5I_{\Delta n}$) ← p188-192

PART 5 : Declaration

1. I/We CERTIFY that the said works do not impair the safety of the existing installation, that the said works, as far as it is reasonably practical to determine, have been designed, constructed, inspected and tested in accordance with BS 7671:2001 (IEE Wiring Regulations), amended to .... and that the said works, to the best of my/our knowledge and belief, at the time of my/our inspection, complied with Chapter 13 of BS 7671:2001

← p87-88, p197-199

2. Electrical Inspector: ....   3. Signature: ....

   For and on behalf of: ....   Position: ....

   Address: ....

   Date: ....

   4. An Inspection of this installation is recommended after _____ years

NAPIT membership Number: Expiry Date:

Copyright NAPIT @ November 2004

ELECTRICAL CONCEPTS

Electricity

Michael Faraday discovered how to make electricity in 1831 when he plunged a bar magnet into a coil wire and thus generated a wave of electricity. He later found that, by rotating a copper plate between the poles of a magnet, power could be taken from the axis to the rim of the disk.

This system of holding the coil of wire stationary while varying the strength of the magnetic field is used in power stations. However, the bar magnet is replaced by a rotating electromagnet and the coils are arranged so that the windings are cut by the magnetic field as the magnet rotates.

Electrical charge

The basis of electrical energy is electric charge. There are two kinds of electric charge: positive and negative.

All materials consist of tiny particles called atoms. Materials which consist of the same kind of atom are called elements. The simplest atom is that of hydrogen. If we could see it, it would resemble the Earth with the Moon orbiting around it.

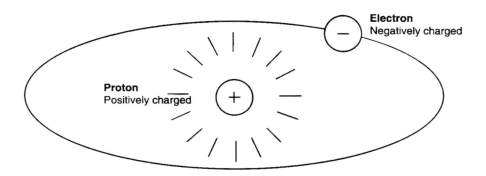

The two types of electric charge play a fundamental role in the atomic structure of matter.

Every atom consists of a nucleus of positively charged particles (called protons), around which negatively charged particles (called electrons) are spinning. Normally, the number of electrons in an atom equals the number of protons in the nucleus, and the atom is then said to be balanced.

If most of the atoms in the piece of material are electrically balanced, the material is said to be electrically neutral, or uncharged. Sometimes, material loses some of the electrons which belong to its atoms so that it contains more protons than electrons. It is then positively charged. Alternatively, an object may gain more electrons than are needed to balance all the protons in the atoms. The object is then negatively charged.

Electric charges share with magnetic poles the property of exerting forces upon one another. It must be remembered that like charges repel one another and unlike charges attract one another. This rule applies to the charges of the electron and proton.

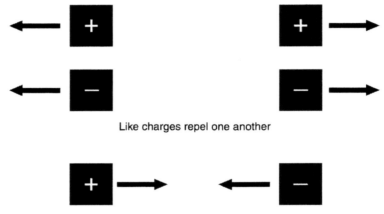

Like charges repel one another

Unlike charges attract one another

Since the nucleus is positive and the electron is negative, the electron is bound in orbit and is normally prevented from flying off because of the attraction between the two particles. The number of orbiting electrons in a given atom depends on the type of element.

The principle that like charges repel and unlike charges attract can be used to force electrons to move in the same general direction and thus to produce current flow.

This is achieved by an external negative charge (which is no more than a point with an abundance of electrons) placed at one end of a conductor and an external positive charge (a point with a deficit of electrons) placed at the other. This causes free electrons to flow towards the positive end, since electrons are negatively charged and unlike charges attract. Further, the external negative charge repels free electrons into the material so that there is ordered directional electron flow from negative to positive, as illustrated.

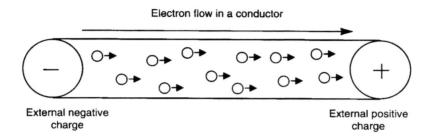

In conductive materials, the electrons not tightly linked to the atoms can be dislodged from their orbit by the force from an electric power supply. In this situation some of the negatively charged electrons are free to move and flow towards the positive terminal of the power supply.

Conductors and insulators

Materials in which the outer electrons are not tightly bound in the atoms and that can easily be dislodged to produce free electrons are called conductors (for example, copper and aluminium). Conversely, in some materials, the orbiting electrons are so tightly bound that they cannot easily be encouraged to break away from their orbits. These materials are called insulators. Insulators have virtually no free electrons available to form an electric current. Examples of insulators are plastics and ceramics.

Types of conductors and insulators

The most common types of conductors found in electrical installations are:

Copper: found in cable and flex
Brass: found in electrical accessories, such as terminal blocks
Nichrome: found in electric fire elements.

The most common type of insulator used in electrical installations is plastic, of which polyvinyl chloride (PVC), a thermoplastic, is the most widely used. Its wide use is due to its ability to be plasticised, which results in a range of flexible plastics, from rigid to pliable. The softer material is used as insulating covering for electric cables and wiring.

Free electrons

Consider a copper atom illustrated in the following diagram. It has 29 orbiting electrons arranged in four shells.

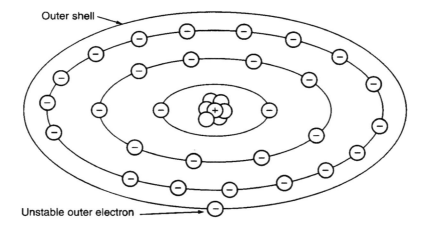

The outer electron, farthest from the nucleus, is only weakly attracted to the positively charged nucleus. This means that it can easily fly off or be dislodged. Electrons dislodged from their orbit can wander at random from atom to atom within the material and so are called free electrons. These free electrons can form the basis of an electric current.

The random movement of electrons

Under normal conditions, free electrons move randomly in a conductor; the effects of temperature cause this movement (see illustration).

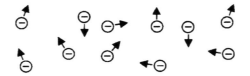

Random electron movement in
a conductor

This movement is usually equal in all directions so that no electrons are given up by the material and none are added. This is not electric current. However, if the free electrons can be encouraged to move in the same general direction within the conductor and can be made to enter and leave it, this flow does constitute an electric current.

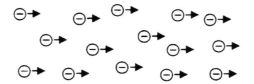

Directional movement – electric
current

Potential difference (electric pressure)

Electrical energy is stored in material when it is either positively or negatively charged. The energy is contained in the charge because, in order to produce it, electrons have to be forced together against the forces of repulsion which exist between them. The amount of charge stored in any given piece of material depends upon how many electrons it has lost or gained. Because electrons are very small, many millions have to be displaced to produce a measurable charge. The extent to which an object is charged is termed 'potential'.

Potential difference is measured in volts.

Voltage is the force behind electricity. It is often referred to as electric pressure and can be readily compared with the water pressure in a plumbing system. The pressure which drives the water is due to the difference of levels between the tank and the tap. The difference in the voltage levels between two points is called the potential difference (p.d.).

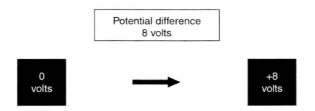

Charge passes across a potential difference only if the two objects are connected by a material which allows electricity to pass through it. Normally, the passage of charge takes the form of a movement of negatively charged electrons but, sometimes, there is a two-way movement of positively and negatively charged particles. When a connection is made between two charged objects, charge passes until the two objects reach the same potential, at which point all movement of charge stops.

Any equipment that works by electricity needs a power supply. Different equipment needs different types of supply. A torch may require two 1.5 V batteries, while an electric shower needs a 230 V AC supply.

The very high voltages that are used on power lines carried by pylons are measured in kilovolts (symbol kV), where one kilovolt is equal to 1,000 V.

Another term for voltage is EMF (electromotive force).

Electric current (amperes)

When charge moves from one place to another, an electric current is said to flow. Electric current is always regarded as flowing from the more positively charged object to the more negatively charged object.

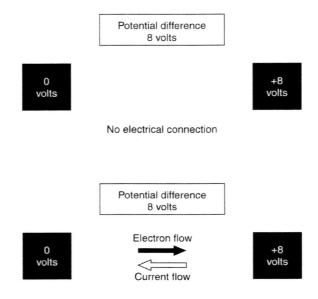

Electrical connection between charges

In a conventional circuit, an electric current (unlike water) will only flow if it can return to its source. The route it takes is known as a circuit. If you break a circuit by cutting a wire that forms that circuit, the current stops.

The unit of current is the ampere, and this is measured in amps (A). An immersion heater takes around 12 A, while the electronic components in a boiler may take only 1 milliamp (mA), which is one thousandth of an amp.

Note: Voltage appears across components and current flows through them.

Resistance

If electric current is like a flow of water, the path it flows along, which is the electrical circuit, can be likened to a heating system with obstacles in its path, such as radiators, that reduce the flow of water in the system.

In electrical circuits even the circuit conductors provide some degree of resistance to the flow of current. This is why voltage is always needed to push the current around the circuit to overcome the resistance.

Consider any piece of equipment – for example, a 110 V electric shaver. If it is accidentally connected to a 230 V supply system, too much current will flow through it and it may burn out. The solution would be to fit some kind of resistance in the circuit to limit the current. This is what happens when electric shavers are made to work on either 230/110 V systems.

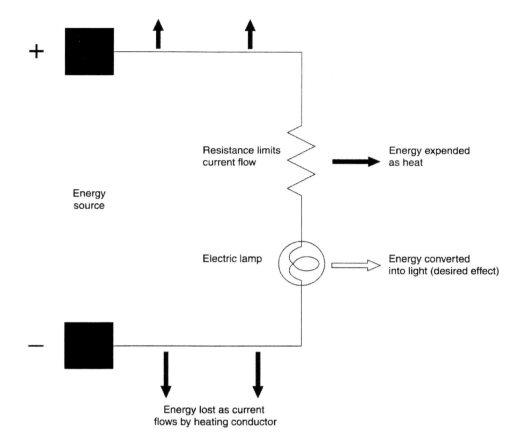

When an electric current meets resistance in a circuit, it generates heat by working its way through the conductor. This heating effect can be put to a practical use – for example, when resistance elements are used in electric fires and also in the filament of a lamp, where the heat produced is enough to make the filament white-hot.

Resistance can cause problems in a circuit, since overheating is a major cause of electrical breakdowns and can give rise to fire risks. If, when terminating conductors into electrical accessories or flexible cords into plug-tops, the terminal screw is not tightened sufficiently, the resulting high-resistance connection will get hot and will damage the accessory. Another factor to consider is the length of cable: the greater the length of run, the greater the resistance, and hence the greater the heat created. Care needs to be taken to ensure that any heat created in cables due to applied load and conductor resistance does not damage the insulation.

The resistance of a component or part of a circuit is measured in ohms, and the omega symbol (Ω) is used to represent it.

Typical values of resistance used in electrical installation and maintenance work are:

ohms 1 k   =   1,000 Ω or 1 kilohm

      1 M   =   1,000,000 Ω or 1 megohm

Understanding electricity

To gain an understanding of electricity, think of it as a plumbing system:

- the height of the tank is voltage (volts)
- the water in the pipework is current (amps)
- the tap is resistance (ohms).

For a given voltage:

- the more you open the tap (less resistance) the more the water flows (more current)
- if you close the tap (more resistance) less water flows (less current).

From this simple analogy it can be seen that:

VOLTS push **AMPS** through **OHMS**

If the water from the tap flows over a water wheel, the speed of the wheel would represent electrical power (watts). The more water that flows, the faster the wheel would turn and therefore the more power.

Relationships (Ohm's law)

If the resistance of a circuit is high, a high voltage is required to push the current round the circuit. When the voltage falls and the resistance of the circuit remains the same, there is less current. From this it is evident that the voltage, current and resistance in a circuit are related to each other. This relationship is known as Ohm's law, after George Ohm, a German physicist.

Expressed mathematically:

$$I = \frac{V}{R} \quad \text{current, in amps} = \frac{\text{EMF (in volts)}}{\text{resistance } (\Omega)}$$

alternatively, $V = I \times R$ or $R = \frac{V}{I}$

In terms of a simple circuit of a battery and flashlamp bulb, if the battery has an EMF of 12 V and the bulb a resistance of 4 ohms, a current of 3 A will flow when the switch is closed.

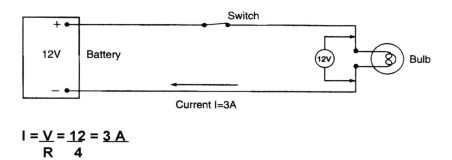

$$I = \frac{V}{R} = \frac{12}{4} = 3\,A$$

Series circuits

In the previous example we were concerned with a simple direct current circuit containing only a battery and lamp. Some circuits contain a number of items connected in series – that is to say, the same current passes through each item, in sequence.

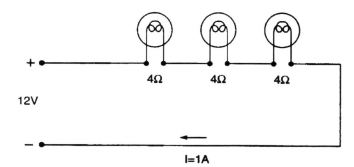

In the above circuit the same current passes through each lamp in turn. If each lamp has a resistance of 4 ohms, the total resistance of the circuit would be 12 ohms. With an EMF of 12 V, it is obvious (applying Ohm's law) that a current of 1 A would flow, and that this 1 A current would be common to all three lamps.

The total EMF of 12 V would be distributed across the total load of all three lamps. If all the lamps were the same (with the same current and power rating factors), each lamp would, in effect, have 4 V across its terminals.

This can be proved by applying Ohm's law, expressed as follows:

Voltage across lamp $\quad V = I \times R$
$\quad\quad\quad\quad\quad\quad\quad\quad\quad 4\,V = 1\,A \times 4\,\Omega$

If the lamps did not all have the same resistance, different voltages would be developed across the terminals of each lamp. For example:

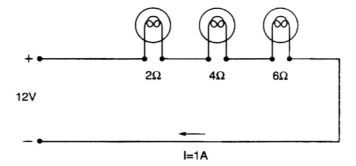

The total resistance of the circuit is still 12 ohms, and a common current of 1 A will flow through each of the lamps in turn. However (from Ohm's law) we can see that the voltage across each lamp would be different. For example:

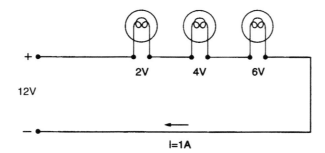

Potential difference (p.d.)

The voltage across each of the lamps is the potential difference (p.d.) required to sustain the common current (in this case of 1 A).

If a similar circuit is devised, but the lamps are replaced with resistors in series, different voltages can be taken at different points. For example:

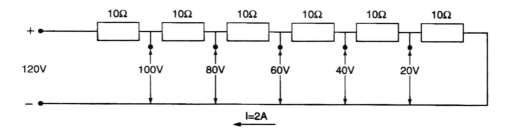

The total resistance of this circuit is 6 x 10 ohms = 60 ohms. If the input voltage is 120 V, a current of 2 A will flow through the circuit. The total voltage of 120 V will be 'divided' across all six resistors. Since they all have the same value (10 ohms), the 'voltage drop' or 'potential difference' across each would be 20 V and the voltage, measured with respect to the common negative, would vary in 20 V steps.

This type of circuit is called a 'potential divider' network and is commonly used where different or varying voltages are required.

Variable resistance

Where a varying voltage is required, this can be achieved by placing a variable resistor in the circuit. For example:

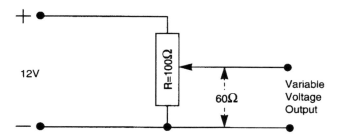

In this type of circuit, the output voltage (with no load connected) is 'tapped off' according to the position of the wiper arm. With an input current of 0.12 A flowing, 7.2 V will be tapped across the upper portion of the resistor, leaving 4.8 V across the output terminals. This type of circuit could be used to provide a reference voltage for, say, a temperature control circuit.

Parallel circuits

Elements in an electrical circuit may also be connected 'in parallel' – that is, the same EMF is applied to each element but the current flowing in each will vary, depending on its resistance. The greater the resistance, the less the current in each element.

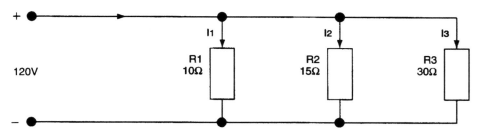

The effective resistance of such a circuit is given by the formula:

$$\frac{1}{R_t} = \frac{1}{R_1} + \frac{1}{R_2} + \frac{1}{R_3} \text{ etc.}$$

$$\frac{1}{R_t} = \frac{1}{10} + \frac{1}{15} + \frac{1}{30}$$

$$\frac{1}{R_t} = \frac{6 + 4 + 2}{60}$$

$$\frac{1}{R_t} = \frac{1}{5} \quad \therefore R_t = \frac{5}{1} = 5\,\Omega$$

Note: R_t = Total resistance.

The current flowing through each element would be:

$$I_1 = \frac{V}{R_1} = \frac{120}{10} = 12\,A \qquad I_2 = \frac{V}{R_2} = \frac{120}{15} = 8\,A \qquad I_3 = \frac{V}{R_3} = \frac{120}{30} = 4\,A$$

If these were added together the total current (It) flowing through the circuit would be:

It = $I_1 + I_2 + I_3$

It = 12 A + 8 A + 4 A

It = 24 A

which confirms the solution above.

Note: It = Total current.

In practice, many devices are connected in parallel and it is important to know the extent of the current flowing in each.

Combined series and parallel circuits

Many circuits include combinations of series and parallel circuits. An example is shown below:

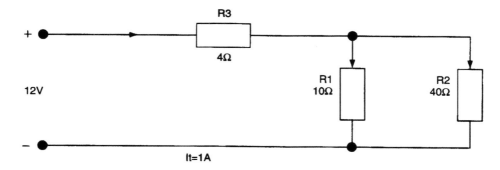

The effective resistance and actual current flow can be determined by employing the principles and formulae discussed previously.

The resistance of the parallel element is determined by the formula:

$$\frac{1}{R_t} = \frac{1}{R_1} + \frac{1}{R_2}$$

$$= \frac{1}{10} + \frac{1}{40}$$

$$= \frac{4+1}{40}$$

$$= \frac{5}{40}$$

$$\therefore R_t = \frac{40}{5} = 8\ \Omega \qquad\qquad \therefore R_t = \underline{8\ \Omega}$$

The parallel element is effectively in series with R_3 so that the effective resistance of the entire circuit is:

4 + 8 = 12 Ω

From Ohm's law, we can determine that the current flowing in the circuit would be 1 ampere. The voltage drop across the series resistance is 4 V, with 8 V across the parallel element.

We can now determine the current flowing in each part of the parallel element, again by the application of Ohm's law:

$$I_1 = \frac{8\ V}{10\ \Omega} \qquad\qquad I_2 = \frac{8\ V}{40\ \Omega}$$

$$I_1 = \underline{0.8\ A} \qquad\qquad I_2 = \underline{0.2\ A}$$

The total current passing through the parallel element is 0.8 + 0.2 = 1 A, confirming the result previously determined.

Electric power

In order to do its work, electricity generates power. Power is the rate at which electrical energy is converted into other kinds of energy, such as heat, light or movement (in the case of electric motors).

The unit of power is the watt, and typical values of power used in electrical circuits are:

Kilowatt = 1,000 watts or 1 kW
Megawatt = 1,000,000 watts or 1 MW

Many electric motors are rated in horsepower – 1 h.p. = 746 watts. Electrical power can be calculated by multiplying the volts by the amps:

Watts (P) = volts (V) x amps (I)

It must be remembered that the formula applies accurately only to DC supplies, but it can be used for rough calculations for AC circuits:

Watts (P) = voltage x current

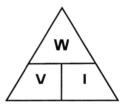

Alternatively, $I = \dfrac{W}{V}$ or $V = \dfrac{W}{I}$

Example: What is the actual current taken by a 3 kW immersion heater, if the supply system is 230 V?

$$I = \dfrac{W}{V} = \qquad I = \dfrac{3,000}{230} \qquad I = 13 \text{ A}$$

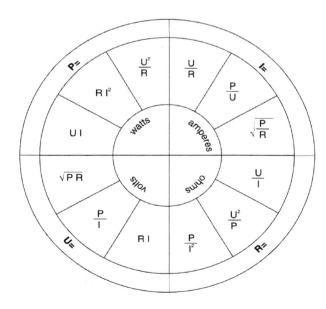

Symbols
U = Voltage in volts
I = Current in amperes
R = Resistance in ohms
P = Power in watts

Types of power supply

There are two types of power supply: alternating current (known as AC) and direct current (known as DC). Alternating current is the type of electricity supplied to domestic, commercial and industrial premises by the local electricity distributor. Direct current is the type of electricity you get from a battery or by using certain components to form a special type of circuit. It is possible to change AC to DC and vice versa.

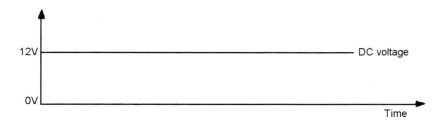

AC theory

All the previous illustrations have related to current which flows in one direction only. The current flowing in a circuit can, however, follow a different pattern (i.e. alternating or AC current), which is shown in the following diagram.

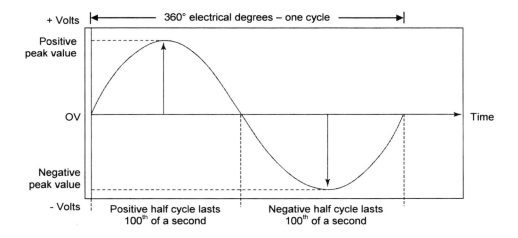

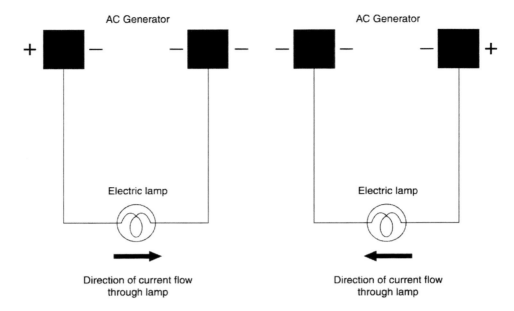

The number of cycles of current flow which occurs in one second is called the frequency of the alternating current. A frequency of 50 Hz is standard for electricity supplies throughout the UK.

Advantages of alternating current

Generators are the only practical means of supplying continuous power. The relative simplicity of the AC generator (or alternator) is one of the advantages of using an AC electrical system. AC, however, has other advantages over DC, which tend to reduce the costs of power production. Among the most important are:

1. some types of electrical apparatus designed to operate from AC supplies are much simpler and easier to maintain than similar apparatus designed to work from DC supplies (in particular, alternating current motors)

2. the voltage of an AC supply can be changed very easily. For example, a low voltage supply can be obtained from a higher voltage supply by means of a transformer (but there is no simple method of changing the voltage of a DC supply)

3. electricity can be distributed more easily and efficiently with an AC supply than with a DC supply. Transformers make it possible to distribute electricity at high voltages. The higher the voltage of transmission, the greater the efficiency due to the fact that they carry less current for a given amount of power and, consequently, the losses or volt drop is reduced.

RMS current and voltage

The root mean square value is the value most commonly used when expressing mains voltage as a value of 230 V.

In order to explain what this means we can use, as an example, a 100 W bulb which glows at a fixed brightness when connected to the mains supply.

The direction of current flowing through the filament plays no part in determining the brightness. This is determined by the power generated within the filament. The current flowing through the filament and, hence, the power level are continually changing as the supply voltage changes. When the mains waveform reaches a peak value, maximum current and hence maximum power are developed in the load. Similarly, as the sine wave voltage reaches zero, no current and therefore no power is generated in the bulb.

If this is the case, why do we not see the brightness vary? The answer is that it would if the frequency of the mains was much lower than 50 Hz. The power and heat are effectively averaged out during each cycle.

To produce the same average power in the load, the peak value must be higher than the RMS value. The relationship between the peak and RMS values has long been established and is illustrated below.

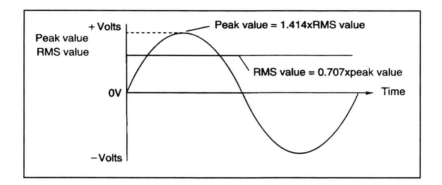

This relationship can be proved by a mathematical process, but this is beyond the scope of this publication.

Electromagnetic induction

Electrical energy is produced in a conductor by the magneto-electric effect created when a conductor is moved in a magnetic field. The creation of a voltage in a conductor by this means is termed 'induction'. We say that a current is induced in a conductor, provided that the conductor forms part of an overall circuit.

It follows that a similar result can be achieved if the conductor is stationary and the magnetic field is moved (i.e. a current is induced in the conductor as a result of the magnetic lines of flux cutting the conductor).

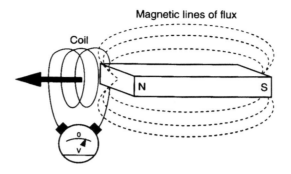

Mutual induction

In the illustration, a permanent magnet provides the magnetic lines of flux to cut the conductor. The permanent magnet could be replaced by an electromagnet. In this case, a current is passed through a coil to produce a magnetic field and the necessary flux to induce a voltage across the conductor. When a current in one coil induces a current in another adjacent coil, this is termed 'mutual induction'.

Transformers

Transformers are inductive devices which use the properties of mutual induction. This means that they rely on a continually changing primary flux to permit an induced voltage to be developed across the secondary. Therefore they can only be used to transform AC voltages; they cannot work at all in pure DC circuits where the primary current is kept constant.

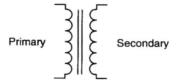

Transformers are used as voltage conversion devices – they change the level of one AC voltage to another, either upwards or downwards. Alternatively, they can be used as isolation devices, allowing two circuits to be coupled without there being a direct electrical connection, as in the case of bathroom shaver units.

If the number of turns on the primary winding equals the number of turns on the secondary, then the induced secondary voltage will equal the applied primary voltage. The transformer is then said to have a 1:1 turns ratio. In practice, losses within the transformer mean that the turns ratio only gives an approximate guide to the primary–secondary voltage relationship.

A turns ratio of 10:1 would produce a voltage across the secondary coil of one tenth of the input voltage (i.e. a 230 V input would produce a 23 V output). Transformers can also be used to 'step up' a voltage. A turns ratio of 1:4 would produce almost 1,000 V output from normal mains input.

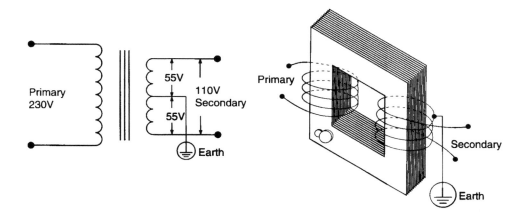

Secondary windings often have a number of graduated 'taps' which provide a variety of outputs. Some transformers include a centre tap in the secondary coil, which is usually earthed. This effectively halves the maximum output voltage relative to earth. In most cases the soft iron core and outer casing are bonded to earth. This type of transformer is used to supply 110 V portable tools on construction sites.

The rating of a transformer

With electrical equipment we must always consider the maximum current that can be carried without exceeding the rating. Transformers are limited in the amount of current they can supply from their secondary windings. If too much current is drawn, the windings get hot, which could melt the winding's insulation and thus burn it out.

Transformers are power-rated in volt-amps (VA). Watts cannot be used to represent power dissipation in a transformer because the voltage and current are not in phase with each other as they are in a pure resistance. By using a volt-amps figure (which is essentially the same, arithmetically speaking), it is relatively simple to calculate the current that can be drawn from a transformer.

When we say a transformer has a particular VA rating we are saying that this is the maximum power that can be drawn from the secondary/secondaries. Consider a mains step-down transformer that has a VA rating of 1 kVA and a single secondary winding of 110 V. What maximum current can safely be drawn from the secondary?

$$\text{Maximum secondary current} = \frac{\text{VA rating}}{\text{Secondary voltage}} = \frac{1000}{110} = \underline{9\ \text{A}}$$

Capacitance

A capacitor (sometimes referred to as a condenser), in its simplest sense, is a device for the temporary storage of electrical energy. It comprises two parallel metal plates, insulated from each other. If a DC voltage is connected across them, one of the plates becomes rich in electrons; the other plate becomes correspondingly poor. In acquiring this charge a current flows, but only for an instant. No sustained direct current can flow between the plates, since they are insulated, one from the other.

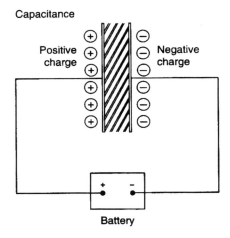

If the DC source is removed, the capacitor will retain its charge until it is discharged through an external circuit.

If an alternating current is fed to a capacitor it will commence to charge on one half-cycle but, as the voltage falls from its peak, it will attempt to discharge and to charge up again (in the opposite direction) on the next half-cycle, and so on. As a result, a capacitor appears to pass current when connected to an AC source, but prevents the passage of DC current.

The larger the area of the plates in a capacitor, the greater the capacitance. In practice, a capacitor is made from two thin sheets of metal foil, insulated by waxed paper, mica or similar material known as a dielectric.

The unit of capacitance is the farad. In practice this is far too big, and the micro-farad (one millionth of a farad) is the unit in common use. This is sometimes written as µF.

REGULATIONS AND STANDARDS

The regulations and standards which cover electrical installations and practices associated with, Part P of the Building Regulations (England and Wales) defined scope level B can be broken down into categories as indicated.

Statutory

The Health and Safety at Work Act 1974
The Electricity Safety, Quality and Continuity Regulations 2002
The Electrical Equipment (Safety) Regulations 1994
The Electricity at Work Regulations 1989
The Building Standards (Scotland) Regulations 1990
The Provision and Use of Work Equipment Regulations 1998
The Building Regulations 2000 for England and Wales Approved Document P

British Standard and Codes of Practice

BS 7671 Requirements for Electrical Installations (The IEE Wiring Regulations)

Memorandum of Guidance on the Electricity at Work Regulations 1989

HSE Guidance Notes

Appendix 1 Memorandum of Guidance on the Electricity at Work Regulations 1989

The Electricity Safety, Quality and Continuity Regulations 2002

These regulations impose requirements with regard to the installation and use of electric lines and apparatus of suppliers of electricity including the provisions for connections with earth.

The regulations also specify that the voltage at the supply terminals shall be no greater than 10% above or 6% below the declared voltage of 230 V for single-phase supplies or 400 V for three-phase supplies.

The Low Voltage Electrical Equipment (Safety) Regulations 1994

These regulations apply to AC equipment operating at voltages above 50 V and below 1,000 V. They also apply to DC equipment operating above 75 V and below 1500 V.

Electrical equipment, together with its component parts, should be made in such a way as to ensure that it can be safely and properly assembled and constructed.

The equipment should be designed and manufactured so as to ensure protection against hazards, providing that equipment is used in application for which it was made and is adequately maintained.

These regulations cover domestic electrical equipment and electrical equipment used in the workplace.

The regulations are intended to establish a single market in safe electrical equipment and provide a high level of protection for consumers throughout the EEC. In the UK these regulations also cover second-hand equipment.

All new electrical equipment that is intended for supply in the UK since 9th January 1995 must fully comply with the requirements of these regulations and as such carry the CE marking.

CE marking is a visible declaration by a manufacturer or his representative that the electrical equipment concerned, fully meets the requirements of these regulations.

The Provision and Use of Work Equipment Regulations (PUWER) 1998

Require employers to ensure that all work equipment is suitable for purpose, is properly maintained and appropriate training is given.

Approved Document P

Electrical Safety, Approved Document P, is a new part of the Building Regulations 2000 for England and Wales and is effective from 1st January 2005.

The purpose of Part P (as it is commonly known) is to improve the standard of competence of Electrical Installers and try to reduce the numbers of deaths, injuries and fires caused by defective electrical installations.

All electrical work in dwellings (this includes outbuildings, garages, sheds, greenhouses, gardens common shared amenity areas of flats etc.) will need to be notified to the building control department of the local authority before work commences, unless:

- the work is carried out by a prescribed competent person (this individual or company authorised to self-certify their work)
- the proposed work in minor work and is not in a kitchen or bathroom (or other area classed as a special location by BS 7671 IEE regulations)

Minor work is the replacing of electrical accessories, e.g. sockets, switches, etc. It also covers adding sockets or lights etc. to an existing circuit. **It does not cover the Installation of a new circuit.**

All electrical work (including minor work) in kitchens and special locations, e.g:

- bath or shower
- swimming pools
- saunas
- garden lighting and power

will need to be notified to building control or self-certified by an authorised competent person.

The Electricity at Work Regulations 1989

Introduction

These regulations came into force on 1st April 1990. They have been written to reduce the increasing number of accidents involving electricity.

Their purpose is to require precautions to be taken against the risk of death or injury from electricity in work-related activities. The emphasis is on prevention of danger from electric shock, burns, electrical explosion or arcing, or from fire or explosion initiated by electrical energy.

The regulations apply wherever the Health and Safety at Work Act applies, wherever electricity may be encountered and to all persons at work. Areas covered are those associated with the generation, provision, transmission, transformation, rectification, conversion, conduction, distribution, control, storage, measurement or use of electrical energy, for example, from a 400 kV overhead line to a battery-powered torch. Therefore there are no voltage limits to these regulations.

In order to provide guidance on the interpretation of the regulations the Health and Safety Executive (HSE) has produced a memorandum of guidance to assist persons involved in the design, construction, operation or maintenance of electrical systems and equipment. Therefore the assessment of danger, and how the regulations are to be applied to overcome it, will be the continuous responsibility of both the employer and employee working as a team to achieve continual compliance with the regulations.

When persons who design, construct, operate or maintain electrical installations and equipment need advice they should refer to guidance, such as may be found in Codes of Practice or HSE guidance, or they should seek expert advice from persons who have the knowledge and experience to make the right judgements and decisions; the necessary skill and ability to carry them into effect. It must be remembered a little knowledge is usually sufficient to make electrical equipment function but it usually requires a much higher level of knowledge and experience to ensure safety.

The Regulations

The following is a brief summary of the regulations. 1-16

Regulation 1: Citation and commencement

The regulations are entitled The Electricity at Work Regulations 1989 and came into force on 1st April 1990.

Regulation 2: Interpretation

Explains the meaning of the following:

- systems
- electrical equipment
- conductors
- danger
- injury

Regulation 3: Persons on whom duties are imposed by these regulations

It shall be the duty of every employer and self-employed person to comply with the regulations for matters within his control.

It shall be the duty of every employee when at work to:

- co-operate with his employer to enable him to comply with the provisions of the regulations
- to comply with the regulations for matters within his control

Regulation 4: Systems, work activities and protective equipment

This regulation has very wide application and requires all electrical systems to be constructed and maintained in a way so as to prevent danger. This includes the design of the system and correct selection of equipment, as well as regular inspection and maintenance to ensure the continuing safety of the system and associated equipment.

The requirement for maintaining detailed and up-to-date records is also stressed.

System of Construction

It is the responsibility of the designer, installer and tester to meet this obligation.

The electrical inspection and test carried out on the commissioning of an installation is to confirm the installation complies with the designer's intentions and has been constructed, inspected and tested in accordance with BS 7671.

System Maintenance

It is the responsibility of the duty holder to meet this obligation.

This refers to the necessity to monitor the condition of the system throughout its life, by implementing a programme of periodic testing and inspecting.

Records of maintenance, including test results should be kept, thus monitoring the condition of the installation, and the effectiveness of the programme.

With regard to work activities, the overriding consideration is that work should not be carried out on a system unless it is 'dead'. The circuit or equipment to be worked on must be properly isolated by locking-off, labeling, etc. and the circuit proved dead at the point of work before work starts.

Systems with more than one possible supply (for example extractor fans with timer circuits) will need extra care to ensure all relevant circuits are properly isolated. The test instrument used (for example voltage indicator or approved test lamp) to prove the circuit or equipment dead, must itself be proved as functioning correctly, immediately before and after testing. This can be achieved using a known supply point or a voltage proving unit.

The last part of this regulation includes the requirement that protective equipment such as special tools, protective clothing and insulating materials, must be suitable for the purpose, maintained in good condition and be properly used. Examples are insulating gloves and floor mats which are covered by British Standards.

Regulation 5: Strength and capabilities of electrical equipment

All equipment must be selected so that it meets appropriate standards (for example British Standard), operates safely under normal and fault conditions, in order to ensure its ability to withstand thermal, electro-magnetic, electro-chemical or other effects of electrical current which will flow when a system is operating. Equipment must be installed and used in accordance with the instructions supplied.

Regulation 6: Adverse or hazardous environments

This regulation covers the requirement to consider the kinds of adverse conditions where danger would arise if equipment is not constructed and protected to withstand such exposure.

Examples of protection from mechanical damage would be running cables through holes in the centre of a joist, instead of in a slot at the top of the joist where the cable could be struck by a nail when fastening down the floorboards.

In the case of weather, when installing frost thermostats the thermostat would have to be suitable for outdoor use with an enclosure and cable entry gland with a suitable IP rating.

Regulation 7: Insulation, protection and placing of conductors

This regulation requires that all conductors in a system which may give rise to danger to be covered with a suitable insulating material and further protection where necessary e.g. conduit, trunking, etc.

The IEE Wiring Regulations provide information and advice for electrical installations up to 1,000 V AC.

Regulation 8: Earthing of other suitable precautions

This regulation applies to any conductor, other than a circuit conductor, which may become charged with electricity either as a result of use or a fault occurring in the system, for example circuit protective conductors (CPCs).

Typical techniques are:

- Earthed Equipotential Bonding Automatic Disconnection (EEBAD). Earthing enables detection of earth faults and the supply to the faulty circuit or equipment to be cut-off automatically by the operation of the circuit protective device

- Equipotential Bonding. All exposed and extraneous conductive parts are inter-connected so that not dangerous potentials can exist between these parts

- Double insulation

- Use of reduced voltage. 110 V centre tapped transformers with 55 V to earth, or extra low voltage (maximum values of extra low voltage are 50 V AC or 120 V DC)

- Use of RCDs, to provide supplementary protection in addition to fuse and circuit breakers

- Separated or isolating transformers (found in bathroom shaver sockets and the supply system for whirlpool baths)

Regulation 9: Integrity of referenced conductors

This refers to the maintenance of the integrity of the earth and/or neutral conductors, for example, not inserting fuses, MCBs or switches into conductors connected to earth.

Regulation 10: Connections

This regulation requires all joints and connections to be both mechanically and electrically suitable for their use.

It therefore covers connections to terminals of plugs, socket outlets, fused spur units, junction boxes and appliances. Therefore all terminations of cables and flexible cords shall be such that the correct amount of insulation is removed, the connections are tight and verified by a tug on the conductor, and that no stress or strain is placed on the conductor or cable itself. This can be achieved by using cord grips in plug tops or appliances and by clipping cables around entries into junction boxes.

Regulation 11: Means of protection from excess of current

To meet the requirements of this regulation, the means of protection is likely to be fuses or circuit-breakers to guard against overload and fault current, the IEE Regulations provide guidance on this subject. This type of electrical protection must be correctly selected, installed and maintained.

Regulation 12: Means for cutting off the supply

This regulation covers two separate functions: cutting-off the supply and isolation.

Devices used for cutting-off the supply are usually switches which must be capable of disconnecting the supply to equipment under normal operating or fault conditions.

They should be clearly marked to show their relationship to the equipment they control unless that function is obvious to persons who need to operate them.

Devices used for isolation are usually switches which have the capability of establishing, when operated, an air gap with sufficient clearance distances to ensure that there is no way in which the isolation gap can fail electrically.

The position of the contacts or other means of isolation should either be externally visible or clearly and reliably indicated.

Typical isolating devices would be a double-pole switch in a consumer unit and in the case of an appliance a 13 A plug and socket.

Both devices for switching and isolating should be positioned so that there is ease of access and operation and the area adjacent is kept free from obstructions.

Regulation 13: Precautions for work on equipment made dead

This regulation states the requirements for such precautions should be effective in preventing electrical equipment from becoming charged with electricity in such a way that would give rise to danger.

The procedures for making equipment dead will usually involve switching off the isolating device and locking it off. If such facilities are not available, the removal of fuses or links which have to be kept in safe keeping can provide an alternate and secure arrangement if proper control procedures are used.

Regulation 14: Working on or near live conductors

This regulation states the requirement for working on or near equipment which has not been isolated and proved dead.

A typical example of live work would be live testing, for example the use of a suitable voltage indicator or multimeter on mains power.

The factors that have to be considered in deciding if it was justifiable for live work to proceed would include:

- when it is not practical to carry out the work with the conductors dead, for example in the case of measuring voltage.

When working on or near conductors which are live, suitable precautions would be the:

- use of people who are properly trained and competent to work on live equipment safely
- provision of adequate information about the live conductors involved, the associated electrical system and the possible risks
- use of suitable tools including insulated tools, equipment and protective clothing
- use of suitable insulated barriers or screens
- The use of suitable test instruments and test probes (GS38)

Testing to establish whether electrical conductors are live or dead should always be done on the assumption they are live until such time as they have been proved dead.

When using test instruments or voltage indicators for this purpose, they should be proved immediately before and after use on a known supply or proving unit. It must be remembered that although live testing may be justified it does not follow that such justification can be made for the repair work to be carried out live. It should be carried out with the conductors safely isolated.

Regulation 15: Working space access and lighting

The purpose of this regulation is to ensure that sufficient space, access and adequate illumination is provided while persons are working on, at or near electrical equipment in order that they may work safely.

Regulation 16: Persons to be competent to prevent danger and injury

The object of this regulation is to ensure that persons are not placed at risk due to a lack of skills on the part of themselves or others in dealing with electrical equipment and the work associated with it.

The requirements are that persons must possess sufficient technical knowledge or experience or be supervised.

For the purpose of this regulation technical knowledge or experience may include an:

- adequate knowledge of electricity
- adequate experience of electrical work
- adequate understanding of the system to be worked on and practical experience of that system
- understanding of the hazards which may arise and the precautions which need to be taken during work on a system
- ability to recognise at all times if it is safe to continue to work.

PLAN AND STYLE OF REGULATIONS BS 7671

The 16th Edition is based on the international regulations produced by the International Electrotechnical Commission (IEC). It is the aim of the IEC to eventually have a common set of wiring regulations.

Format

The regulations are divided into seven parts with seven appendices. This format is illustrated overleaf.

Numbering

Each part of the regulations is numbered consecutively, being identified by the first number of each group of digits. The parts are divided into chapters, identified by the second digit and each chapter is split into sections, identified by the third digit. After the first group of three digits, the digits separated by hyphens identify the regulation itself.

Example 1

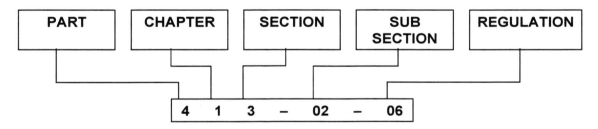

Example 2

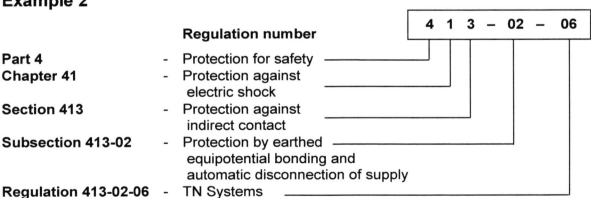

	Regulation number	
Part 4	- Protection for safety	
Chapter 41	- Protection against electric shock	
Section 413	- Protection against indirect contact	
Subsection 413-02	- Protection by earthed equipotential bonding and automatic disconnection of supply	
Regulation 413-02-06	- TN Systems	

Relevance to Statutory Regulations and British Standards

The IEE Wiring Regulations have the status of a British Standard. The full title is *British Standard 7671: 2001 Requirements for Electrical Installation (The IEE Wiring Regulations)*. The following statutory regulations recognise the IEE Wiring Regulations as a code of good practice:

- Health and Safety at Work Act 1974
- Electricity at Work Regulations 1989
- Electricity Safety, Quality and Continuity Regulations 2002

Note: On 1 January 1995, CENELEC, the European electrical standards body, implemented the following changes in nominal voltage:

240 V ± 6% to 230 V +10%/−6%
415 V ± 6% to 400 V +10%/−6%

This is to harmonise low voltage supplies within Europe. To complete this harmonisation, it is proposed on 1 January 2008 to make the final adjustment to 230 V +10%/−10%.

BS 7671: 2001 was issued on 1 June 2001 and came into effect from 1 January 2002.

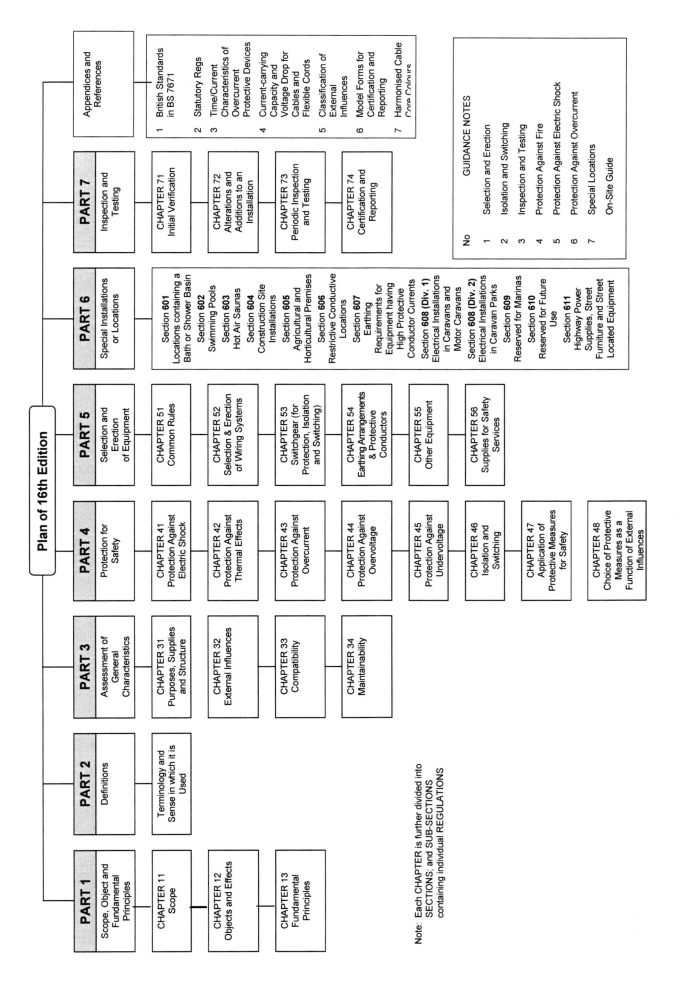

SCOPE, OBJECT AND FUNDAMENTAL PRINCIPLES *(Part 1)*

Scope *(Chapter 11)*

The regulations apply to electrical installations covered by Part P of the Building Regulations:

- residential premises

Requirements are included for:

- voltages up to and including 1000 V AC or 1500 V DC
- any wiring systems and cables not covered by an appliance standard
- all consumer installations external to buildings
- alterations and additions to installations and the subsequently affected parts of existing installations

Voltage Ranges

The regulations cover installations with the following operating voltages:

- **Extra low voltage**: 0 V to 50 V AC or 120 V ripple-free DC, whether between conductors or to earth
- **Low voltage**: Exceeding extra low voltage to 1000 V AC or 1500 V DC between conductors, or 600 V AC or 900 V DC between conductors to earth

Voltage Bands

Band I

- Installations that for operational reasons have their voltage limited, e.g. alarms, controls, telecommunications, bells, etc.
- Extra low voltage

Band II

- The supplies to electrical and industrial installations
- Low voltage (Band II voltages do not exceed 1000 V RMS AC or 1500 V DC)

Relationship with Statutory Regulations

The IEE Wiring Regulations BS 7671 are non-statutory, although they may be used as evidence in a court of law to prove compliance with a statutory regulation (see BS 7671, Appendix 2).

Object and Effects *(Chapter 12)*

BS 7671 contains the rules for the design and erection of electrical installations that ensures their safe operation and correct function.

BS 7671 gives the technical requirements to enable compliance with the fundamental principles of Chapter 13

- Part 3 Assessment of general characteristics
- Part 4 Protection for safety
- Part 5 Selection and erection of equipment
- Part 6 Special locations
- Part 7 Inspection and testing

Fundamental Principles *(Chapter 13)*

Protection for Safety *(130-01)*

This section contains the general requirements for the safety of:

- persons
- livestock
- property

Against the risks which could arise in the normal use of electrical installations.

Risk of injury could be from:

- shock
- excessive temperatures likely to cause burns or fires, etc.
- mechanical movement of electrically actuated equipment
- explosion

Protection Against Electric Shock *(130-02)*

Persons and livestock should be protected, so far as is reasonably practicable, against the dangers arising from contact with live parts.

Direct contact protection to be provided by:

- preventing current flowing through persons or livestock
- OR, limiting the value of current that can flow through a body to below the shock current

Indirect contact protection to be provided by:

- preventing current flowing through persons or livestock
- OR, limiting the value of current that can flow through a body to below the shock current
- OR, automatic disconnection when a fault arises that is likely to cause a current in excess of the shock current to flow through a body in contact with exposed conductive parts

The application of equipotential bonding is a **vitally** important safety requirement.

Protection Against Thermal Effects *(130-03)*

Installations should be designed so that the risks of fire due to high temperature or arcing are reduced so far as is reasonably practicable.

During the normal use of equipment, the risk of burns to persons or livestock should be minimised.

Persons, fixed equipment or fixed materials next to electrical equipment should be protected from the harmful effects of heat or thermal radiation emitting from the electrical equipment, especially:

- combustion, ignition, or degradation of materials
- burns
- unsafe functioning of equipment

Electrical equipment should not create a fire hazard to adjacent materials.

Overcurrent Protection *(130-04)*

Persons and livestock to be protected from injury, and property from damage, due to high temperatures and electromechanical stresses caused by overcurrent.

Fault Current Protection *(130-05)*

All conductors to be capable of carrying any fault currents without achieving excessive temperatures.

Overvoltage Protection *(130-06)*

Persons and livestock to be protected from injury, and property from damage, due to faults between live parts of circuits supplied at different voltages and any overvoltages that may arise due to switching or atmospheric conditions.

Additions and Alterations *(130-07)*

Before an addition or alteration is made to an existing installation it must be verified that the ratings and condition of the existing equipment (including the distributor's equipment) is suitably adequate for any increase in loadings and that the earthing and bonding arrangements are satisfactory.

Design *(131)*

An electrical installation must provide for the:

- protection of persons, livestock and property
- proper functioning of the electrical installation

Supply Characteristics *(131-02)*

The supply details shall be determined by calculation, measurement, enquiry or inspection.

1. Nature of current (AC and /or DC)
2. Purpose and number of conductors.

    For AC:
    - Phase conductor(s)
    - Neutral conductor
    - Protective conductor

    For DC:
    - Equivalent to those conductors opposite
    - Outer, middle, earthed live conductors
    - Protective conductor

3. Values:
    - nominal voltage and tolerances
    - nominal frequency and tolerances
    - maximum current allowable
    - earth fault loop impedance
    - protective measures in the supply (e.g. earthed neutral)
    - the distributor's particular requirements

Nature of Demand *(131-03)*

The circuits required for lighting, power, control, signalling, communication and IT equipment, etc. are to be determined from the following information:

- location of points of power demand
- loadings
- daily and annual power demand variations
- special conditions

Environmental Conditions *(131-05)*

Equipment should be constructed or protected to prevent danger from exposure to the weather, corrosive atmospheres or other adverse conditions. Equipment in areas with a high risk of fire or explosion must be constructed or protected and any other special precautions taken to prevent danger.

Cross-sectional Area of Conductors *(131-06)*

Conductor size is to be decided according to:

- admissible maximum temperature
- volt drop
- electro-mechanical stresses occurring due to short circuit or earth fault currents
- other likely mechanical stresses on the conductors
- the maximum impedance values for operation of the short circuit or earth fault protection

Type of Wiring and Method of Installation *(131-07)*

This is to be determined with consideration of the following:

- nature of location
- structure supporting the wiring
- accessibility of the wiring to persons and livestock
- voltage
- electro-mechanical stresses due to short circuit and earth fault currents
- any other stresses to which the wiring could be exposed during the erection or use of the installation (e.g. thermal, mechanical and fire-related stresses)

Protective Equipment *(131-08)*

Protective equipment has to be selected for:

- overcurrent (overload and short circuit)
- earth fault current
- overvoltage

Protective devices must operate at values of current, voltage and time with respect to the circuit characteristics and the possibilities of **danger**.

Disconnecting Devices *(131-10)*

A means of disconnecting the supply must be provided to enable maintenance, testing, fault detection and repair to be carried out on installations, circuits or individual items of apparatus.

Prevention of Mutual Detrimental Influence *(131-11)*

No mutual detrimental influences are to occur between individual electrical installations and non-electrical installations. Electromagnetic interference should also be taken into account.

Accessibility of Electrical Equipment *(131-12)*

There must be:

- enough space for the original installation and any future replacement of individual items
- accessibility to allow for operation, inspection, testing, maintenance and repair

Protective Devices and Switches *(131-13)*

- single pole protective devices or switches to be in the phase conductor only

Isolation and Switching *(131-14)*

There must be effective means, suitably placed for ready operation, to enable all voltage to be cut off from every installation to prevent or remove danger.

To prevent danger, every fixed electric motor must have an efficient means of switching off that is readily accessible, easily operated and suitably located.

Selection of Electrical Equipment *(132)*

Every item of equipment shall comply with EN (European Norm), HD (Harmonised Document) or National Standard implementing the HD. In all other cases, reference to the IEC or an appropriate National Standard is required.

Equipment should be suitable for the nominal voltage and any overvoltage possible. For some equipment, the lowest voltage that could occur should be taken into account.

All items of equipment should be suitable for their design current and any currents likely to flow in abnormal conditions, including the time period for operation of protective devices.

Electrical equipment should be suitable for the frequency of the circuit and any equipment chosen for its power characteristics shall be suitable for the demand.

Equipment must withstand the stresses, environmental conditions and characteristics of its location. Any equipment that is not suitable for its location can only be used if further suitable additional protection is provided.

Electrical equipment itself should not cause any harmful effects on other equipment or disrupt the supply during normal use (this includes any switching operations).

Erection, Verification and Periodic Inspection and Testing (133)

Good workmanship and the correct materials should be used.

The equipment should not be impaired by the erection process.

Conductors should be identified and joints and connections should be electrically and mechanically suitable for use.

Equipment design temperatures should not be exceeded and equipment that could cause high temperatures or arcing should be guarded to minimise any risks of ignition. If the temperature of an item of electrical equipment could cause burns to persons or livestock, that too should be guarded to prevent any accidental contact.

Installations, additions or alterations should be inspected and tested on completion to establish whether or not compliance with BS 7671 has been achieved.

The person carrying out the inspection and test should make recommendations for future periodic inspection and testing as specified in Chapter 73 of BS 7671.

Type of Earthing Arrangement

The type of earthing arrangement(s) to be used must also be determined.

The choice of arrangements may be limited by the characteristics of the energy source and any facilities for earthing.

Nature of Supply *(313)*

The following characteristics should be ascertained for the supply or supplies (by calculation, measurement, enquiry or inspection).

- Nominal voltage(s)
- The nature of current and frequency
- The earth fault loop impedance Z_e of that part of the system external to the installation
- Suitability for the requirement of the installation, including maximum demand
- Type and rating of the overcurrent protective device at the origin of the installation

Installation Circuit Arrangement *(314)*

Every installation should be divided into circuits as necessary to:

- avoid danger and inconvenience in the event of a fault, and
- facilitate safe operation, inspection, testing and maintenance

The number of final circuits required in an installation and the number of points supplied by a final circuit shall be arranged to comply with the requirements for overcurrent protection (Chapter 43), isolation and switching (Chapter 46), and current-carrying capacity of conductors (Chapter 52).

Each final circuit must be connected to a separate way in a distribution board, and the wiring of each final circuit should be electrically separate from every other final circuit.

External Influences *(Chapter 32)*

The chapter dealing with the external influences likely to affect the design and safe operation of the installation is not yet at a stage for adoption as a basis for national regulations. Appendix 5 of the regulations contains some useful information on the subject. The codes are utilised in Chapter 52 of the regulations.

Maintainability *(Chapter 34)*

Maintainability is also a very important factor to consider when deciding on the design of an installation.

Only then can the regulations be applied so that:

- Any periodic inspection, testing, maintenance and repairs likely to be necessary during the intended life can be readily and safely carried out and
- The protective measures for safety remain effective
- The reliability of equipment is appropriate to the intended life

THE BUILDING REGULATIONS 2000 FOR ENGLAND AND WALES

The regulations are divided into fourteen parts. Each part deals with an aspect of building design and construction as listed:

- A   Structure
- B   Fire safety
- C   Site preparation and resistance to contaminants and moisture
- D   Toxic substances
- E   Resistance to the passage of sound
- F   Ventilation
- G   Hygiene
- H   Drainage and waste disposal
- J   Combustion appliances and fuel storage systems
- K   Protection from falling, collision and impact
- L   Conservation of fuel and power
- M   Access to and use of buildings
- N   Glazing – safety in relation to impact, opening and cleaning
- P   Electrical safety

The requirements within each part are often called 'functional requirements' and are written in terms of what is:

- reasonable
- adequate
- appropriate

Any of the requirements which are relevant to a particular part of building work must be complied with.

The parts which effect electrical installation work are:

- A   Structure
- B   Fire safety
- E   Resistance to the passage of sound
- F   Ventilation
- L   Conservation of fuel and power
- M   Access to and use of buildings
- P   Electrical safety

LIGHTING, POWER CIRCUITS AND HEATING SYSTEM CONTROLS

Drawings and Circuit Diagrams

A technical drawing or diagram is simply a means of conveying information more easily or clearly than can be expressed in words. In the electrical industry drawings and diagrams are used in different forms. Most frequently used are:

- block diagrams
- circuit diagrams
- wiring diagrams
- layout diagrams
- 'as fitted' drawings
- detail drawing

Block Diagrams

The various items are represented by a square or rectangle clearly labelled to indicate its purpose.

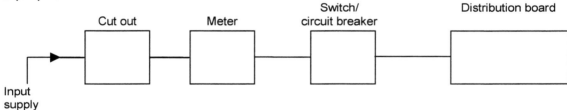

Circuit Diagrams

A circuit diagram makes use of special symbols to represent pieces of equipment or apparatus and to show clearly how a circuit works. It may not indicate the most convenient way of wiring the circuit but it will show the electrical relationship between the various circuit elements.

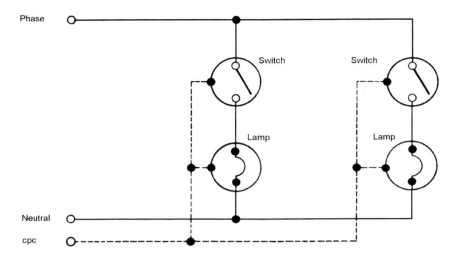

Lighting circuit

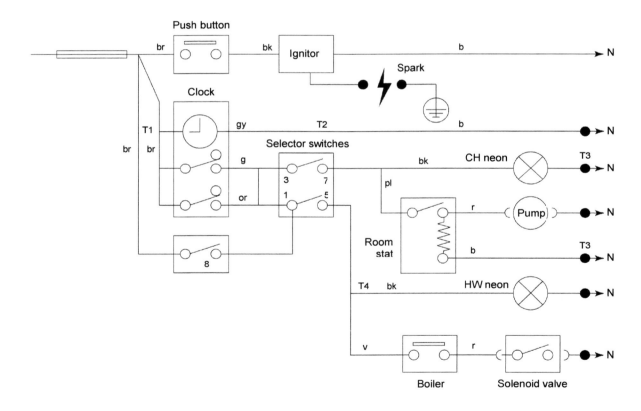

Heating control circuit

Wiring Diagrams

Wiring diagrams give sufficient information for the connection of a circuit. In some respects, wiring diagrams may be more detailed than circuit diagrams but they do not necessarily give any indication of how the equipment concerned operates.

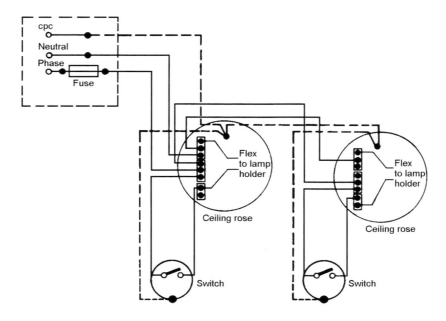

Wiring diagram for loop-in ceiling rose lighting system

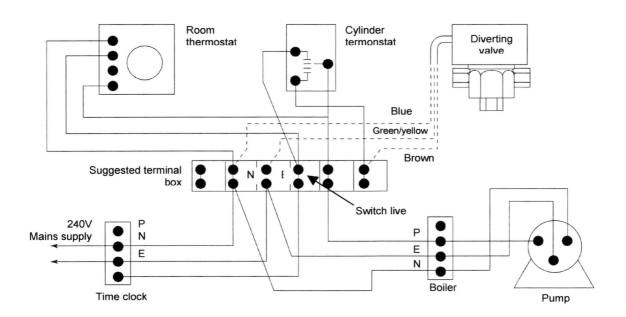

Wiring diagram for a heating circuit

Cabling Diagrams

Cabling diagrams shows the cables required for the interconnection between components of an installation and will specify the type of cable, number of cores, earth connections, etc.

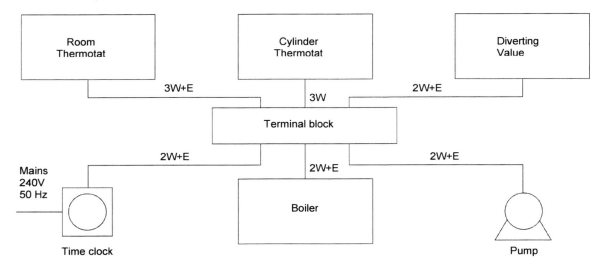

Cabling diagram

Layout Diagrams

Layout diagrams show the physical layout of the installation in pictorial or symbolic form.

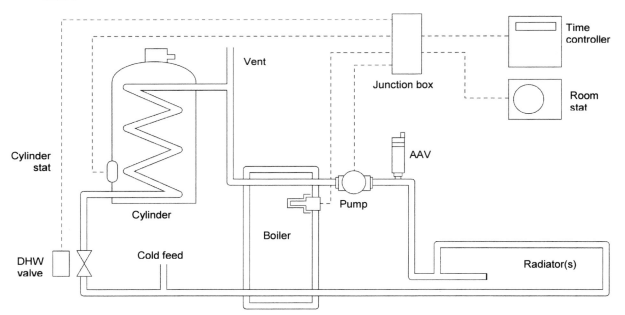

Layout diagram

Electrical systems depend on there being a complete circuit from the source of supply to the apparatus and back again. Every circuit must start from one pole of the supply and return to another pole (e.g. from phase to neutral). In an installation supplied by a supply authority, the origin of the installation is at the output terminals of the supply company's meter (or double pole switch in the case of modern domestic supply systems).

Circuits are interrupted so that the power may be controlled by:

- switches
- fuses
- contactors and relays
- time switches
- thermostats.

Planning Final Circuits

A final circuit is that part of an installation between an item of current consuming equipment and the fuse or circuit breaker at the consumer unit. It is, in effect, the point where load is applied to a system. The design of final circuits is very important and every precaution must be taken to ensure they do not constitute a hazard to the user.

There are many factors that have to be considered in the design of final circuits. Some are purely electrical – others are concerned with location.

Consider first the electrical factors. The rating of the cables, switches and protective devices must be decided together with the form and size of circuit protective conductor and the type of protection to be used.

Physical factors such as the distance from the supply to the load, which could introduce a voltage drop, must also be considered. The temperature and the proximity to other cables and thermal insulation are all factors that would require larger cables to be selected for the given load.

LIGHTING CIRCUITS

One-way Lighting Circuits

The simplest circuit consists of a pair of wires from the mains terminals supplying a lamp. In this circuit there must be a switch that (if single-pole) must be situated in the phase conductor.

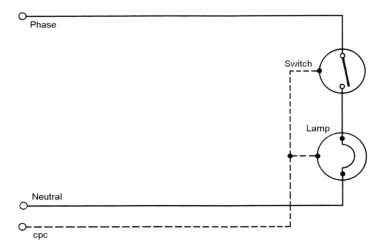

Two-way Lighting Circuits

A two-way lighting circuit is often used on staircases so that one can switch off the downstairs light from upstairs, or vice versa.

In this circuit, the switches can have two positions, either of which can light the lamp. Suppose switch A is in the upper position and switch B is in the lower position, as illustrated, there is no circuit so the lamp is out.

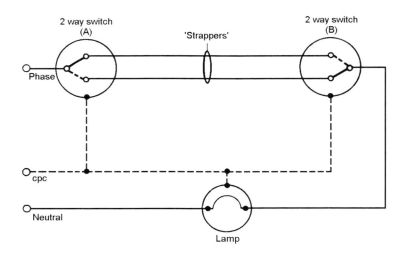

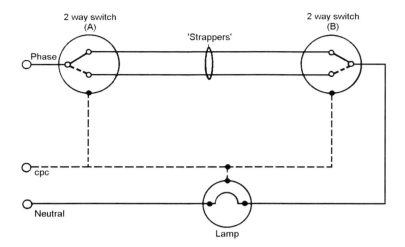

If switch B is operated, a circuit is established and the lamp lights. Now if switch A is operated, the lamp goes out. The two wires between switches A and B are called 'strappers'.

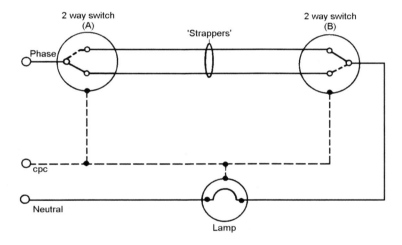

Intermediate Lighting Circuits

When it is necessary to control the lighting in a long corridor or staircase with several landings, it is desirable to arrange for lights to be switched on and off at several points. In this case, intermediate switches should be used.

If, as illustrated, the two wires between switches A and B (the strapping wires) were reversed, a circuit would be established and the lamp would light.

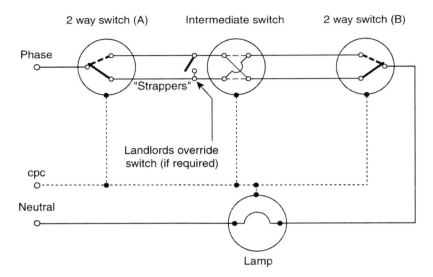

The intermediate switch carries out this reversal of the strapping wires. Any number of intermediate switches may be installed.

Converting a One-way Lighting Circuit To Two-way Switching

It is sometimes necessary to modify lighting circuits that are controlled from one point.

Consider an existing one-way circuit such as the one from the beginning of this section and shown again here.

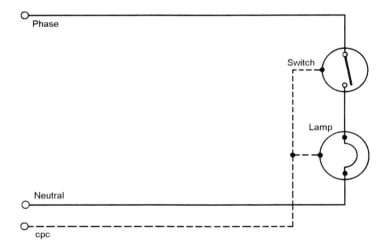

To convert it to a conventional two-way circuit, either the switch feed, or the switch wire, must be removed and replaced by a conductor going to another switch position. This alteration must be made either at the lamp or at the consumer unit.

This can be avoided by using the circuit shown below. Two-way switch control can be achieved by replacing the original switch with a two-way switch connected as illustrated and by running three new wires to a two-way switch at the new control position.

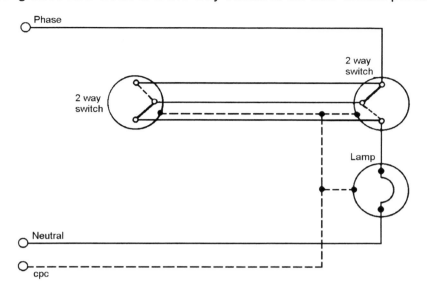

The Joint Box Method

For two lights that are independently controlled, the circuit would be as illustrated.

1. Joint box showing phase, neutral and circuit protective conductor (cpc).

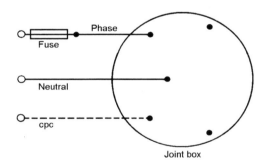

2. Switch connected showing switch feed, switch wire and cpc.

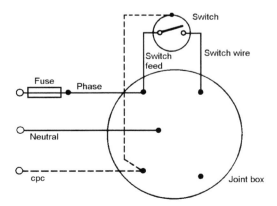

3. Light connected showing switch wire, neutral and cpc.

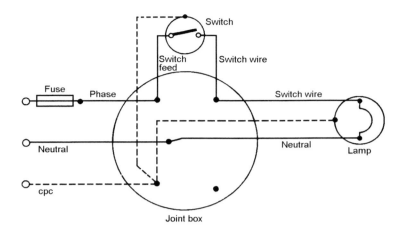

4. Additional light connected showing all switch feeds, switch wires, neutrals and cpc.

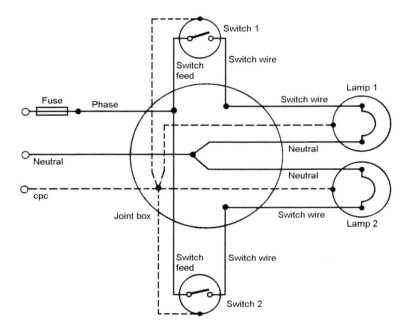

The 'Loop-in' Method

The most common system of wiring final sub-circuits is the loop-in system where all connections are made at the electrical accessories.

For simplicity, all wiring diagrams have to show the basic circuit wiring necessary for the circuit. The loop-in system of wiring at the ceiling rose would be as follows.

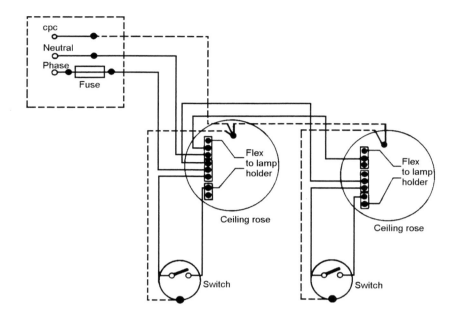

Lampholders

Lampholders must not be connected to any circuit where the value of the overcurrent protective device exceeds those given below (reference should be made to IEE Regulations, Table 55B).

Maximum rating of overcurrent protection devices of circuits		
Type of lampholder		**Max. rating (Amps)**
Small Bayonet Cap	B15	6
Bayonet Cap	B22	16
Small Edison Screw	E14	6
Edison Screw	E27	16

Lighting Points

For each fixed lighting point, one of the following must be used:

- ceiling rose to BS 67
- luminaire supporting coupler to BS 6972 or BS 7001
- batten lampholder to BS 7895, BS EN 60238 or BS EN 61184
- luminaire designed to be connected directly to the circuit wiring
- suitable socket outlet
- a connection unit to BS 5733 or BS 1363-4

When luminaire supporting couplers (LSC) are required, they must not be used for the connection of any other equipment – they are designed specifically for mechanical support and the electrical connection of the luminaire. If the LSC has a protective conductor contact, it cannot be used on a SELV system.

Lighting accessories or luminaires must be controlled by a switch or switches to BS 3676, BS 5518 or by a suitable automatic control. They must also be appropriate, where necessary, for the control of discharge lighting circuits.

Ceiling roses must not be installed in any circuit operating at voltages in excess of 250 V and must not be used for the attachment of more than one flexible cord, unless specially designed for multiple pendants.

POWER CIRCUITS

13 Amp Socket Outlet Circuits

The 13 A socket outlet system is based on the principle of 'diversity of use'.

Types of final circuit using BS 1363 socket outlets and fused connection units are:

- ring circuits
- radial circuits

These two types of circuit are described in detail on the following pages.

Final circuits using BS 1363 socket outlets and connection units (Table 8A of the IEE On-Site Guide) are shown below.

Circuit	Minimum conductor size (PVC insulated, copper conductors)	Copper conductor MI cable	Type and rating of overcurrent device	Maximum floor area
Ring	2.5 mm²	1.5 mm²	30 A or 32 A	100 m²
Radial	4 mm²	2.5 mm²	30 A or 32 A	75 m²
Radial	2.5 mm²	1.5 mm²	20 A	50 m²

Ring Circuits

In this system, the phase, neutral and circuit protective conductors are connected to their respective terminals at the consumer unit and loop into each socket in turn. They then return to their consumer unit terminals, forming a 'ring'.

Each 13 A socket has two connections back to the mains – each capable of carrying 13 A at least.

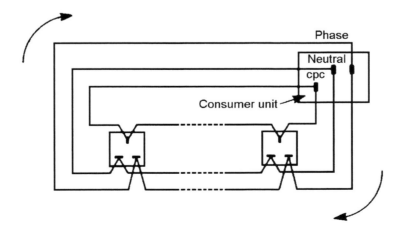

The requirements for ring circuits are as follows:

- an unlimited number of outlets may be provided. (Each socket outlet of a twin or multiple socket is to be regarded as one outlet.)

- the floor area served by a single 30 A or 32 A ring circuit must not exceed 100 m²

- consideration must be given to the loading of the circuit, especially kitchens which may require a separate circuit

- when more than one ring circuit is installed in the same premises, the outlets installed should be reasonably shared amongst the ring circuits so that the assessed load is balanced

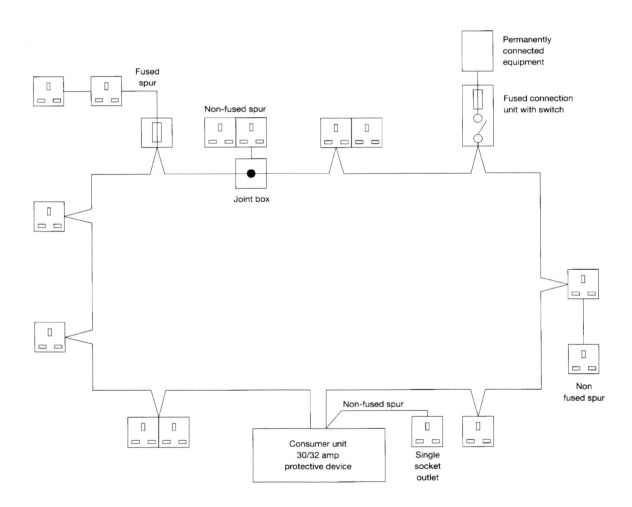

A typical ring circuit

Spurs

The total number of fused spurs is unlimited, but the number of non-fused spurs must not exceed the total number of socket outlets and any stationary equipment connected directly to the circuit.

Non-fused Spurs

A non-fused spur may supply only one single or one twin socket outlet or one item of permanently connected equipment.

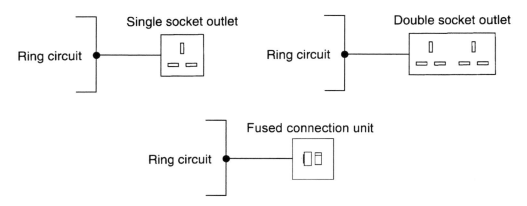

Note: The cable size of non-fused spurs should not be less than that of the ring circuit.

The spurs are connected to the circuit as follows.

    a) At the terminal of accessories on the ring

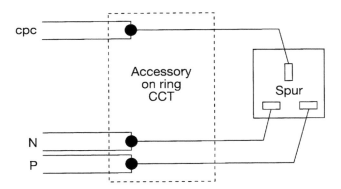

b) At a joint box

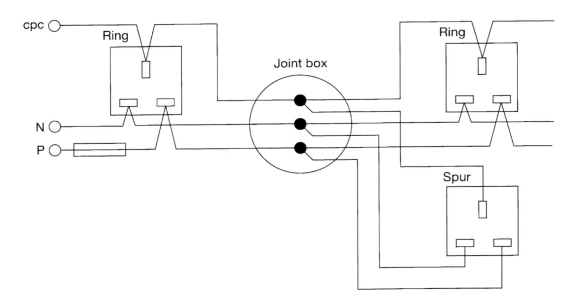

c) At the origin of the circuit in the distribution board

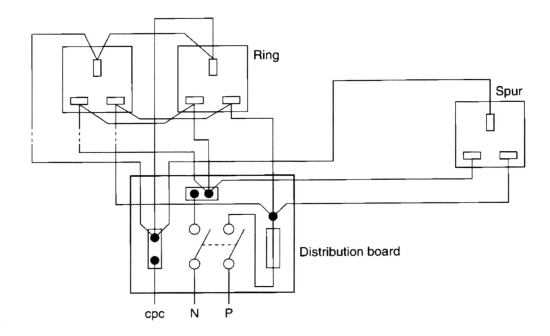

Fused Spurs

A fused spur is connected to a circuit through a fused connection unit. The fuse incorporated should be related to the current carrying capacity of the cable used for the spur, but should not exceed 13 A.

When outlets are wired from a fused spur the minimum size of conductor is:

- 1.5 mm² for PVC or thermosetting insulated cables with copper conductors

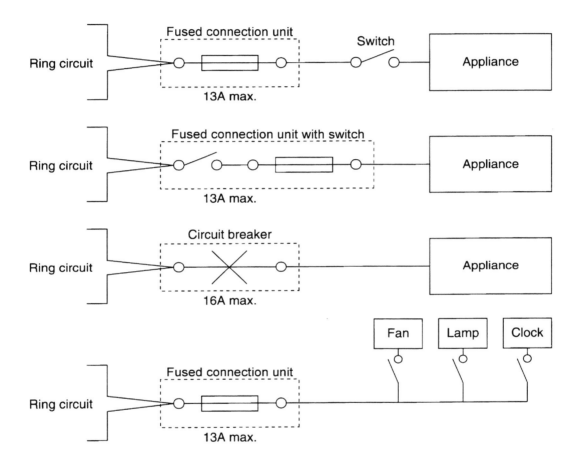

Note: The cable size for fused spur is dependent on the magnitude of the connected load.

Radial Circuits

Radial circuits also make use of 13 A sockets (BS 1363) but the circuit is not wired in the form of a ring.

An unlimited number of outlets may be supplied, but the floor area which may be served by the outlets is limited to either 50 m² or 75 m² depending on the size and type of the cable used and size of overcurrent protection afforded.

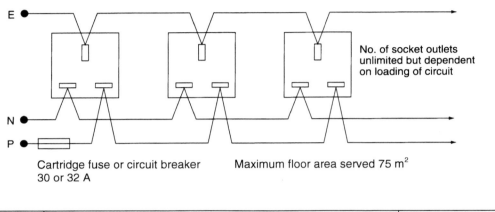

Cartridge fuse or circuit breaker 30 or 32 A

Maximum floor area served 75 m²

Minimum cable size	Copper conductor PVC or thermosetting insulation	4 mm²

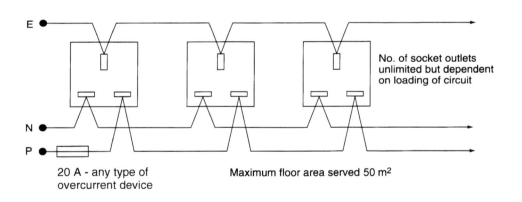

20 A - any type of overcurrent device

Maximum floor area served 50 m²

Minimum cable size	Copper conductor PVC or thermosetting insulation	2.5 mm²

Typical radial circuits using BS 1363 socket outlets

Fixed Equipment

Where immersion heaters are installed in storage tanks with a capacity in excess of 15 litres, or a comprehensive space heating installation is to be installed (for example, electric fires or storage radiators), separate circuits should be provided for each heater.

Immersion Heaters

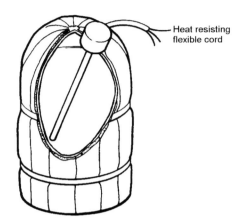

Cable protection – damage by heat

With a permanently installed hot water system, the heater is placed in a cylinder or tank. Hot water is drawn off at sink, basin or bath.

Storage cylinders and tank systems all use electric heating elements immersed in the water, allowing for an efficient transfer of heat. They are generally thermostatically controlled.

The immersion heater must be wired on a separate radial final circuit.

The flexible cord used to make the final connection between the circuit and the immersion heater may reach quite a high temperature, especially if covered by lagging materials, therefore a heat-resistant (HR) flexible cord should be used.

Electric Shower Circuits

There are various models available and these vary considerably in size, shape and capacity.

A typical final circuit diagram is illustrated below.

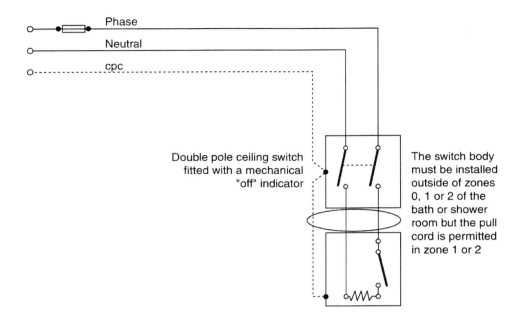

Dual Immersion Heater Circuits

It is possible to fit two immersion heaters with thermostats into the same cylinder to achieve either a full, or a part cylinder of hot water.

A typical dual immersion heater circuit diagram is illustrated.

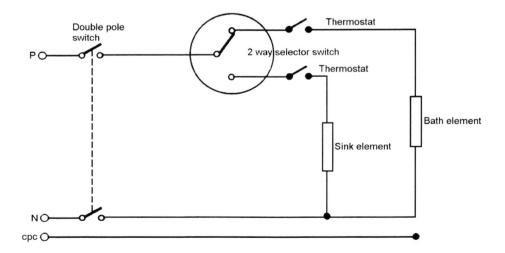

Cooker Circuits

A circuit supplying a cooking appliance must include a control switch or cooker control unit (which may include a socket outlet). The rating of the circuit should be determined by referring to Table 1B of the IEE On-Site Guide.

There are no special regulations relating to the installation of cooking appliances. However, when installing cooking appliances, IEE Wiring Regulations (BS 7671) Section 476 must be complied with. In particular, Regulation 476-03-04 requires that fixed or stationary appliances (connected other than by a plug and socket to 537-05-04) must have means of interrupting the supply on load.

The device used could be incorporated in the appliance itself or, if separate, must be in a readily accessible position placed so as not to put the operator in danger. Where two or more appliances are installed in the same room, one switch could be used to control all the appliances.

The method of wiring for a standard cooker is shown below. A cable is run direct from the consumer unit to a cooker control unit or double switch, then to a cooker outlet to which the cooker is then connected.

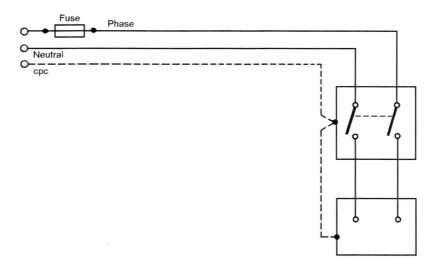

Wiring diagram for an electric cooker circuit

When wiring a 'split level' cooker, a cable is run from the consumer unit to a double-pole isolating switch. Subject to electrical loading and accessibility of the switch from both appliances (the IEE On-Site Guide recommends a maximum of 2 m from the appliance to the switch), the two units may be wired from one switch. A typical layout is shown on the next page.

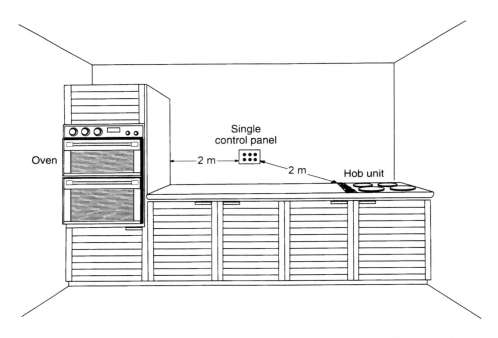

Control of fixed and stationary appliances in domestic premises

In fitted kitchens that incorporate built-in appliances, switched fused connection units could be installed above the worktop, feeding 13 A socket outlets installed below the worktop. Alternatively, the socket outlets could be controlled from a multi-gang grid switch assembly (suitably rated and protected).

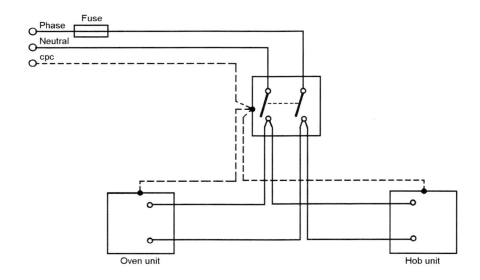

Wiring diagram for split level electric cooker

HEATING SYSTEM CONTROLS

The Elements of a Control System

Three elements are necessary for a control system:

1. a sensor
2. a controller
3. the device being controlled.

Electrical Symbols

All types of diagram may make use of standard symbols and it is important that you should be familiar with these. Some of the most common are listed below:

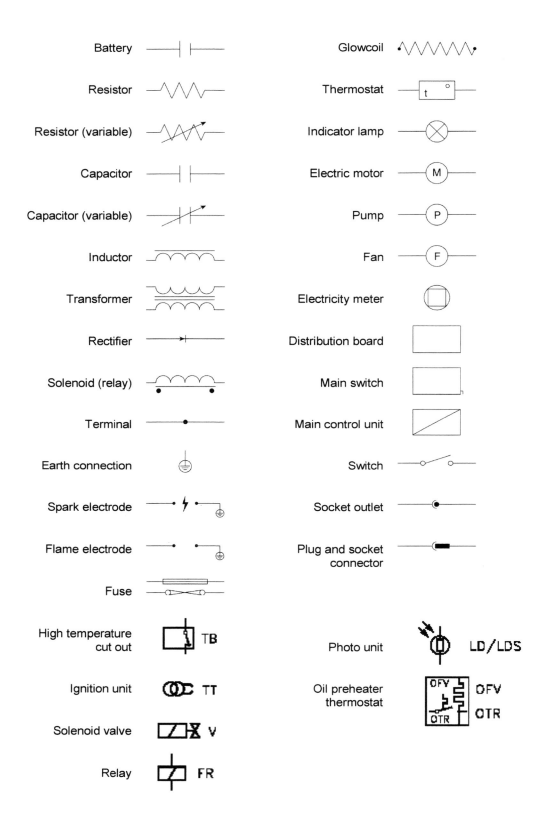

Typical control systems

The following are examples of controls which are often found on gas and oil fired heating systems.

Zone valve controlled gravity system

This is designed to provide independent temperature control of both heating and hot water circuits in pumped heating and gravity domestic hot water heating installations. Time control may be provided by a time switch or programmer.

Note: This type of controls (and wet) system does not usually achieve compliance with the requirements of the Building Regulations for new installations, but in certain circumstances may be able to be used to compliantly upgrade an existing installation. Compliance time control for heating and hot water should be able to be independently controlled, for example, via a two-channel time switch or programmer or by the use of a single channel time switch and a programmable room thermostat.

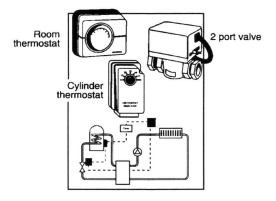

The operation of this type of system is as follows:

Heating only – on a demand for heat from the room thermostat, the pump and boiler are switched on. The zone valve in the domestic hot water primary remains closed.

Hot water only – on a demand for heat from the cylinder thermostat, the valve is energised open. Just before the valve reaches the fully open position, the auxiliary switch closes and switches on the boiler only.

Heating and hot water – when both thermostats demand heat, the pump and boiler are switched on and the domestic hot water valve is open.

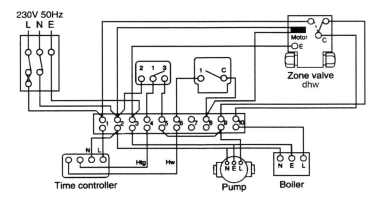

Twin zone valve control system

This is designed to provide independent temperature control of heating and domestic hot water circuits in fully pumped central heating installations. Time control may be provided by a simple time switch or a programmer.

Note: This type of controls (and wet) system can be utilised as part of the means of achieving compliance with the requirements of the Building Regulations for both new installations and as an upgrade to an existing installation. Compliance time control for heating and hot water should be able to be independently controlled, for example, via a two-channel time switch or programmer or by the use of a single channel time switch and a programmable room thermostat.

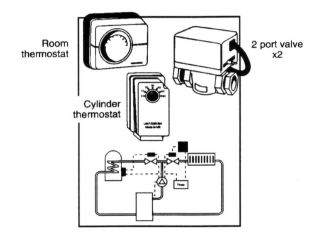

On a demand for heat from either thermostat, the respective heating or domestic hot water zone valve will be energised open. Just before the valve reaches its fully open position, the auxiliary switch will be closed and switch on both pump and boiler. When both thermostats are satisfied the valves are closed and the pump and boiler switched off.

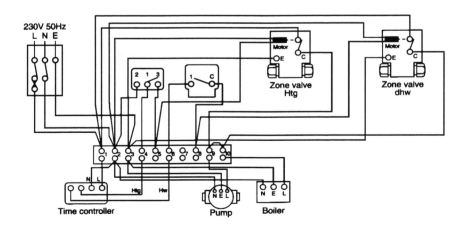

Hot water priority diverter valve system

The hot water priority diverter valve system is designed to provide independent temperature control of both heating and domestic hot water circuits in fully pumped central heating installations.

Note: This type of controls (and wet) system can be utilised as part of the means of achieving compliance with the requirements of the Building Regulations for both new installations and as an upgrade to an existing installation. Compliance time control for heating and hot water should be able to be independently controlled, for example, via a two-channel time switch or programmer or by the use of a single channel time switch and a programmable room thermostat.

This system uses a two position diverter valve which is normally installed to give priority to the domestic hot water circuit. Because this is a priority control system, it should not be used when there is likely to be a high hot water demand during the heating season which could lead to the space temperature dropping below comfort level. The situation is likely to occur, for example, in large family dwellings or in poorly insulated properties. Time control may be provided by a time switch or mini-programmer.

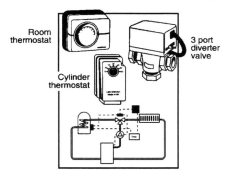

The operation of this type of system is as follows:

Heating only requirement – on a demand for heat from the room thermostat, the valve is energised, so that the hot water port is closed, and the heating port is open and the pump and boiler switched on.

Hot water only requirement – on a demand for heat from the cylinder thermostat, the pump and the boiler are switched on.

Heating and hot water – when wired and installed to give hot water priority, the heating will remain off until the hot water is up to the required temperature.

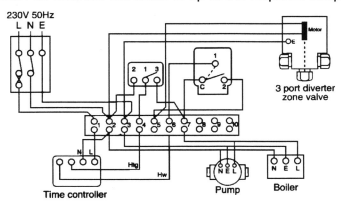

Full control diverter valve system

This is designed to provide independent temperature control of both heating and domestic hot water circuits in fully pumped central heating installations. Time control may be provided by a simple time switch or a programmer.

Note: This type of controls (and wet) system can be utilised as part of the means of achieving compliance with the requirements of the Building Regulations for both new installations and as an upgrade to an existing installation. Compliance time control for heating and hot water should be able to be independently controlled, for example, via a two-channel time switch or programmer or by the use of a single channel time switch and a programmable room thermostat.

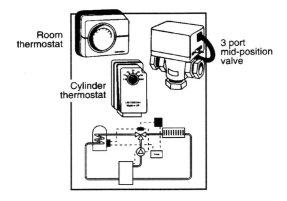

This system operates as follows:

Heating only requirement – on a demand for heat from the room thermostat, the valve motor is energised so that the central heating port only is opened and the pump and boiler switch on.

Hot water requirement – on a demand for heat from the cylinder thermostat, the valve remains open to domestic hot water only and the pump and boiler switch on.

Heating and hot water – when both thermostats demand heat, the valve plug is positioned to allow both ports to be open and the pump and boiler switch on. When neither thermostat is demanding heat, the pump and boiler are off. The valve remains in the last position of operation whilst the time control is in the ON position.

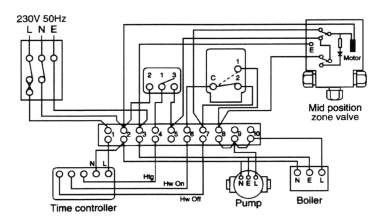

EQUIPMENT AND ACCESSORIES

The Distributor's Equipment

Most of our electricity is generated at large, remote power stations and then transmitted the relatively long distances into our towns and cities. At a certain point, it becomes the responsibility of the local distributor, who through his network (underground and service cables, equipment, etc.) will distribute it to our premises. It is the service cable that is terminated at the cut out, in a mutually agreed position within the premises.

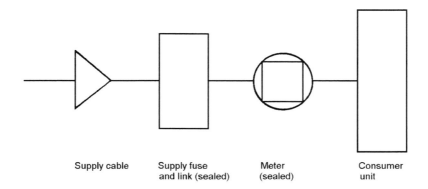

Supply cable   Supply fuse and link (sealed)   Meter (sealed)   Consumer unit

At the service termination point, the service cable is connected to a service fuse (normally 80 A or 100 A) that is designed to protect the distributor's cables and equipment in the event of a fault occurring on the consumer's premises. The supply should be in accordance with the Electricity Safety, Quality and Continuity Regulations 2002.

Connections are installed by the local distributor from the service fuse to the meter, which records the amount of electricity consumed in units or kilowatt-hours (kWh). The service fuse and meter are sealed by the local distributor – and consumers or contractors are not permitted to interfere with this part of the installation.

The Consumer's Installation

The consumer's installation begins at the output terminals of the local distributor's meter or, in the case of a domestic installation, the local distributor's meter or double-pole switch (the switch, if fitted, is on the outgoing side of the meter).

Main switches must be provided so that the installation may be isolated from the supply when alterations or extensions are made.

Consumer Control Units

For domestic installations where the load does not exceed 100 A, a consumer control unit is normally installed.

A consumer control unit constructed with an open back is not acceptable since the enclosure must provide protection to at least IP2X or IPXXB with the top of the enclosure being to IP4X (see Regulation 412-03-01).

Consumer control units are manufactured to BS EN 60439-3 and usually consist of a 60, 80 or 100 A double-pole switch or a residual current device (RCD) and a number of output ways or modules.

An arrangement with eight output ways could be used to supply the following circuits:

- two 32 A ways for two-ring final circuits
- one 32 A way for the cooker circuit
- one 20 A way for the garage power circuit
- one 16 A way for the immersion heater circuit
- two 6 A ways for the lighting circuits
- one spare way

There will be many variations – no two types of installation will have exactly the same requirements.

Modern consumer control units are very versatile – manufacturers produce a wide range of components that can be fitted into the selected enclosure. These include:

- circuit breakers
- combined circuit breaker/RCD (RCBO)
- fuses (cartridge type, see Regulation 533-01-04)
- time switches
- contactors
- transformers

Enclosures are selected for a given module size subject to the needs of the installation.

The main switch could be a double-pole switch or an RCD then a range of circuit breakers or BS 1361 fuses to protect the individual circuits.

Typically the main switch or RCD would occupy two module ways each, and the individual circuit protective devices occupy one module each. Some control devices will take up two or even three module ways (such as contactors, time switches) and allowances should be made at the design stage.

Split Load Consumer Control Units

Split load consumer control units are the popular choice for modern domestic installations. They give sensitive RCD protection to selected circuits, with the unit being specifically designed to suit the installation requirements.

The main switch could be:

- a double-pole switch
- a 30 mA RCD

This would be the consumer's main switch for the installation and would supply a number of output ways for selected circuits that do not require 30 mA RCD protection (such as lighting). It would also supply the 30 mA RCD that protects a further number of output ways for circuits that do require a high level of earth fault protection (such as power).

Therefore, if an earth fault occurs on the ring circuit, the 30 mA RCD would operate leaving the main switch side of the consumer unit energised and the lighting would remain on. This avoids plunging the building into darkness with the additional risk of physical injury.

With a standard consumer unit that has a double-pole main switch, it may be possible to install an RCBO (combined RCD and circuit breaker) within the unit to give RCD protection to an individual circuit.

Both metal-clad or insulated consumer control units are available. If a conduit system is being installed, a metal clad unit is normally used. For sheathed cable systems, an insulated enclosure is more usual.

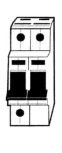
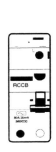

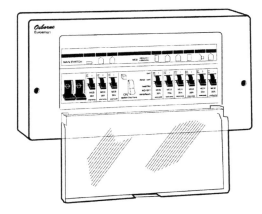

Main Switchgear

Every installation must be controlled by one or more main switches. The main switchgear may consist of a switchfuse or a separate switch and fuses and must be readily accessible to the consumer and as near as possible to the local distributor's intake. The Electricity at Work Regulations state that 'suitable means shall be available for cutting off the supply and isolating the supply to any electrical equipment'.

The IEE Wiring Regulations BS 7671 Regulation 131-14-01 requires that 'effective means, suitably placed for ready operation, shall be provided so that all voltage may be cut off from every installation, from every circuit thereof and from all equipment as may be necessary to prevent or remove danger'.

Control of Separate Installations

Control apparatus must be provided for every section of a consumer's installation. An off-peak supply (for example, one supplying electric storage heaters) is considered to require separate metering and control apparatus in addition to the metering and control apparatus for lighting and other loads.

Where a consumer's installation is split into separately controlled parts, each part must be treated as a separate installation. This applies irrespective of whether the parts are within the same building or in separate detached buildings.

Every installation must be controlled by one or more main switches. The main switchgear may consist of a switchfuse or separate switch and fuses.

Accessories

Plugs and Socket Outlets

Plugs and socket outlets are recognised as being suitable for electrical installations by the IEE Regulations for low voltage circuits (reference should be made to Table 55A of the IEE Regulations).

Plugs and socket outlets for low voltage circuits (including types, ratings and relevant British Standards) are shown below.

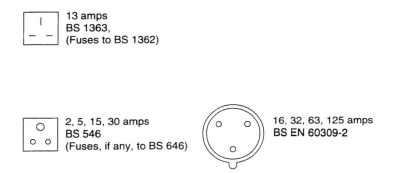

These plug and socket outlets are designed so that it is not possible to engage any pin of the plug into a live contact of a socket outlet whilst any other pin of the plug is exposed (not a requirement for SELV circuits) and the plugs are not capable of being inserted into sockets of systems other than their own.

With the exception of SELV or special circuits having characteristics where danger may arise, all socket outlets must be of the non-reversible type, with a point for the connection of a protective conductor.

Plugs and socket outlets, other than those shown above, may be used on single-phase AC or two-wire DC circuits operating at voltages not exceeding 250 V for the connection of:

    Electric clocks - Clock connector unit incorporating a fuse BS 646 or 1362 not exceeding 3 amperes

    Electric shaver – BS EN 60742 shaver supply unit for use in bath or shower rooms. In other locations, a socket to BS 4573 can be used

On construction sites (but not necessarily in site offices, toilets, etc.) only plugs, sockets and couplers to BS EN 60309-2 must be used.

Where socket outlets are mounted vertically, they should be fixed to a height above floor level or working surface so that the plug and associated flexible cord are not subjected to mechanical damage during insertion, use or withdrawal of the plug.

|M| Approved Document M of the Building Regulations for new installations in new dwellings require outlet points such as sockets, telephone and TV to be installed at a minimum height of 450 mm from finished floor level or a maximum of 1200 mm.

Cable Couplers

Except for SELV circuits or Class II circuits, cable couplers may be used in conjunction with the following types of plug and sockets:

- BS 196
- BS 4491
- BS EN 60309-2
- BS 6991

Cable couplers should be connected so that the plug is on the load side of the installation. They should be non-reversible and have provision for the connection of the protective conductor.

Summary

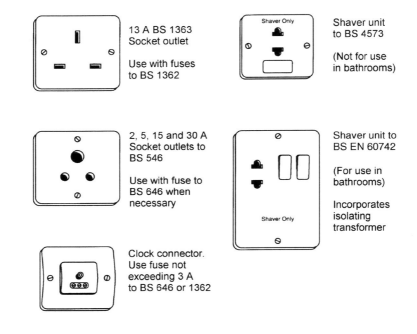

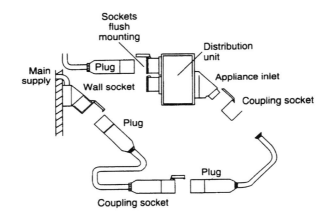

BS EN 60309-2 plugs, socket outlets, cable couplers and inlets

Wiring Accessories

The range of different wiring accessories used for wiring circuits is wide. If we consider, for instance, lighting circuits, you will find ceiling roses connected by a flex to lampholders to form a pendant or batten lampholders for direct fixing to the ceiling and of course, switches for turning the lights on and off. These may be wall or ceiling mounted and in the case of wall switches, may be flush or surface mounted on plastic moulded or metal boxes.

Lampholders

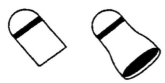

Short skirt   HO protective skirt

Pendants and ceiling rose

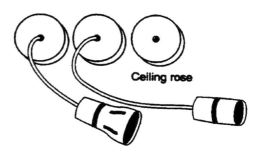

Ceiling rose

HO pendant   Short skirt pendant

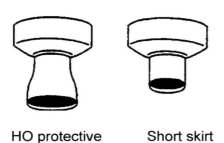

HO protective   Short skirt

Note: HO protective types used in bathrooms.

Plate switches

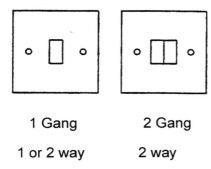

1 Gang   2 Gang

1 or 2 way   2 way

Ceiling switch

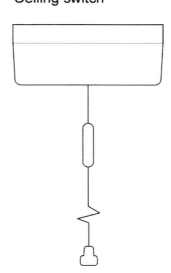

M Approved Document M of the Building Regulations for new installations in new dwellings require lighting switches to be installed at a height from finished floor level no greater than 1200 mm.

For power circuits you will find single or double socket outlets which are available in switched and unswitched formats, again with plastic or metal mounting boxes of the flush or surface type. When control of permanently wired equipment, for example, a central heating boiler or waste disposal unit is required, a switched fused connection unit is required. Switched and unswitched types are available with or without a flex outlet facility.

13 Amp socket outlets   Fused connection (spur) units

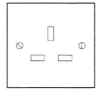 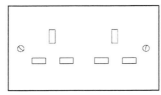 

1 Gang unswitched   2 Gang unswitched   Switched

Double pole switch

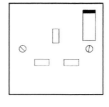 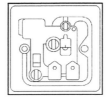 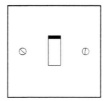 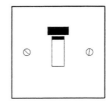 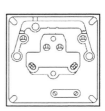

1 Gang switched   Rearview   With pilot light   Rearview

Boxes

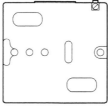 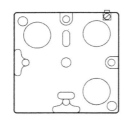 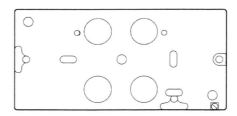

Flush steel 1 gang 17 mm deep   Flush steel 1 gang 27 mm deep   Flush steel double gang 29 mm deep

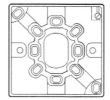 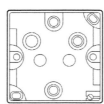 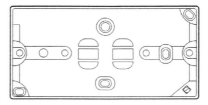

Moulded 1 gang 17 mm deep   Moulded 1 gang 29 mm deep   Moulded double gang 29 mm deep

When there is a need to control individual equipment, for example, electric showers and immersion heaters, a double-pole (isolation) switch should be used which disconnects both the phase and neutral conductors from the supply. These switches may be wall or ceiling mounted, with typical current ratings of 20 amp used for immersion heater circuits, 30 or 45 amp used for shower installations with the size being dependant on the rating of the shower.

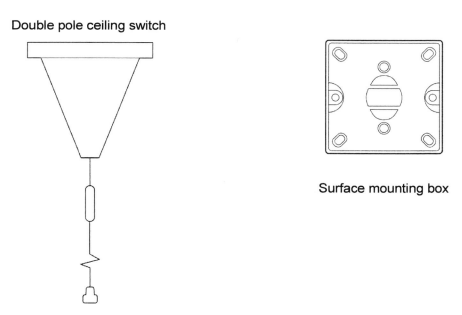

Double pole ceiling switch

Surface mounting box

When wiring some lighting and power circuits it is not always convenient to make the connection in the accessory so junction boxes are used. The most common types are 4, 5 and 6 terminal 20 amp and 3 terminal 30 amp.

Junction boxes

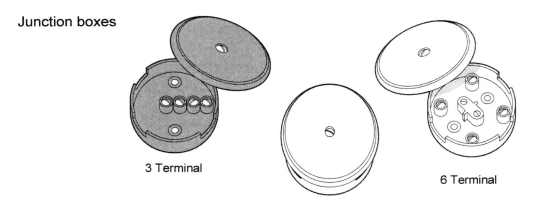

3 Terminal

6 Terminal

Plugs

In order to connect an appliance to a power source a 13 amp plug is generally used which is fitted to the flexible cord of the appliance and is then inserted into the socket outlet. The 13 amp plug has a fuse fitted which is there to protect the appliance and flexible cord against any overload or short circuit conditions. It is very important to fit the correct size fuse, for example a washing machine would require a 13 amp fuse whilst a gas central heating boiler would only require a 3 amp.

The size of fuse most commonly available are 3, 5 and 13 amp, whist 1, 2, 7 and 10 amp types can usually be obtained from manufacturers, to order.

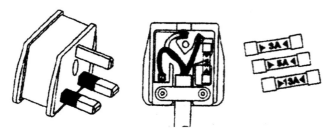

Replacement Fuse Links To BS 1362

3 amp   Red
5 amp   Black
13 amp Brown

Overcurrent Protection Devices

Fuses

The different types of fuse include:

- semi-enclosed fuses, often referred to as rewireable (BS 3036)
- cartridge fuses (BS 1361 and BS 1362)
- high breaking capacity, referred to as HBC (BS 88-2 and BS 88-6)

The advantages and disadvantages of each type are listed below.

Semi-enclosed (or Rewireable Fuses)

Advantages

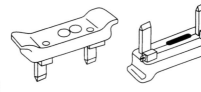

- No mechanical moving parts
- Simple to observe whether element has melted

Disadvantages

- Danger on insertion with fault on installation
- Can be repaired with incorrect size fuse wire
- Element cannot be replaced quickly
- Deteriorate with age
- Lack of discrimination
- Can cause damage in conditions of severe short circuit
- BS 7671 prefers cartridge fuses (see Regulation 533-01-04)

The diameter of copper wires to be used as fuse elements in this type of fuse is given in Table 53A of the IEE Wiring Regulations.

Cartridge Fuses (BS 1361)

The body of the fuse can be either ceramic (low grade) or glass with metal end caps to which the fuse element is connected. The fuse is sometimes filled with silica sand.

Advantages

- Small physical size
- No mechanical moving parts
- Accurate current rating
- Not liable to deterioration

Disadvantages

- Can be replaced with incorrect cartridge
- Not suitable where extremely high fault current may develop (typically 16.5 KA max)
- Can be shorted out by the use of silver foil or wire

HBC Fuses (BS 88)

The barrel of the high breaking capacity (HBC) fuse is made from high-grade ceramic to withstand the mechanical forces of heavy current interruption.

Plated end caps afford good electrical contact.

An accurately machined element, usually made of silver, is shaped to give precise characteristic.

The barrel is filled with quartz silica sand to ensure efficient arc extinction in all conditions of operation.

Some fuses are fitted with an indicator bead that shows when it has blown.

Advantages

- Discriminates between overload currents of short duration (e.g. motor starting) and high fault currents
- Suitable where high levels of PFC could occur (typically 40–80 KA duty rating)
- Consistent in operation
- Reliable

Disadvantage

- Could be replaced with a fuse of incorrect current rating

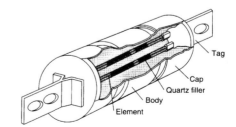

Colour Coding of Fuses

BS 646	
Rating (amps)	Colour
1	Green
2	Yellow
3	Black
5	Red

BS 1361	
Rating (amps)	Colour
5	White
15	Blue
20	Yellow
30	Red
45	Green
60	Purple

BS 1362	
Rating (amps)	Colour
3	Red
13	Brown

Note: The colour for all other ratings is black

Circuit Breakers

Types of circuit breaker to BS EN 60898 (and miniature circuit breaker to BS 3871) include:

- thermal and magnetic
- magnetic hydraulic

Advantages

- Tripping characteristic set during manufacture; cannot be altered
- They will trip for a sustained overload but not for transient overloads
- Faulty circuit is easily identified
- Supply quickly restored
- Tamper proof
- Multiple units available

Disadvantages

- They have mechanically moving parts
- Characteristics affected by ambient temperature

Circuit breakers are now classified according to BS EN 60898. This has replaced BS 3871-1: 1965: the classification of miniature circuit breakers (mcbs). However, the technical data for mcbs has been retained alongside that for circuit breakers since, for the foreseeable future, mcbs will continue to be widely found in existing installations.

BS EN 60898 and BS 3871 instantaneous tripping currents

Type	Ampere			
1	>	$2.7\ I_n$	≤	$4.0\ I_n$
2	>	$4.0\ I_n$	≤	$7.0\ I_n$
3	>	$7.0\ I_n$	≤	$10\ I_n$
4	>	$10\ I_n$	≤	$50\ I_n$
B	>	$3\ I_n$	≤	$5\ I_n$
C	>	$5\ I_n$	≤	$10\ I_n$
D	>	$10\ I_n$	≤	$20\ I_n$

Typical applications for circuit breakers and mcbs

Type	Application
1 B	General domestic and commercial installations with little or no surges
2 3 C	General commercial and industrial installations where fluorescent lighting and small motors produce switching surges
4 D	For use where transformers, industrial welding equipment, x ray machines, etc. where high inrush currents could happen

Duty ratings

Miniature circuit breakers (BS 3871-1)		Circuit breakers (BS EN 60898)	
Category of duty	Prospective current of the test circuit (A)	Icn	Ics
		kA	kA
M 1	1000	1.5	1.5
M 1.5	1500	3.0	3.0
M 3	3000	6.0	6.0
M 4.5	4500	10	7.5
M 6	6000	15	7.5
M 9	9000	20	10.0
		25	12.5

Two rated short circuit capacities – **Icn** and **Ics** – may be quoted for circuit breakers where:

- **Icn** is the rated ultimate short circuit capacity

    This is the maximum fault current the device can safely interrupt, although the device may be damaged and no longer usable

- **Ics** is the rated service short circuit breaking capacity

    This is the maximum level of fault current operation without loss of performance of the circuit breaker

The circuit breaker will be marked with the Icn value (rated short circuit capacity).

The value will appear inside a rectangle without a unit symbol, e.g. $\boxed{10{,}000}$

BS EN 60898 circuit breakers have rated short circuit values of 1.5 kA–25 kA, in practice the smallest rating will be 3 kA. The larger 25 kA values are new and take into account modern current limiting technology. The old BS 3871 and the new BS EN 60898 standard conditions for test and the test duty sequences are entirely different. It is therefore difficult to assess if a 10 kA rating to BS EN 60898 is better or worse than a M9 rating from BS 3871. The user must be aware of these variations when comparing devices manufactured to the different standards.

RESIDUAL CURRENT DEVICES (RCDs)

Residual current devices give protection not only against fire risk but also give adequate protection against shock risk.

Residual Tripping Currents

10 mA — Used in special applications where additional protection is required due to the nature of the installation or location e.g. schools, laboratories, etc.

30 mA — Used to provide supplementary protection to reduce the risk associated with direct contact, e.g. portable equipment used outdoors.

100 mA — Suitable for use against indirect contact shock protection but mainly used for a high level of fire protection or where the earthing requires supplementing with an RCD to achieve compliance with BS 7671, rather than as additional protection for persons. Some manufacturers use this device as the main switch in split load consumer control units.

100 mA time delayed — Installed where total RCD protection is required for an installation to supplement the earthing to achieve compliance with BS 7671 and where 30 mA RCDs are used for supplementary direct contact protection, e.g. main switch of a split load consumer control unit.

300 mA — Provides means to achieve compliance with BS 7671 in areas of poor earth loop impedance, good fire protection.

500 mA — Can be used on large 3 phase installations to improve earth loop values and where local 30 mA RCDs are fitted. Also used for fire protection, required by Regulation 605-10-01 for agricultural installations.

Residual Current Devices

This term refers to a complete range of devices including:

    RCCB - Residual Current Circuit Breaker

    RCBO - Residual Current Breaker with Overcurrent device

    SRCD - Socket outlet with Residual Current Device

    PRCD - Portable Residual Current Device

An RCD is designed to give protection against shock risk and against fire. The basic circuit for a single-phase device is as illustrated.

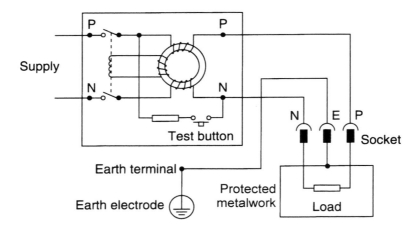

Basis of Operation (Single-phase)

Two conductors pass through, or are wound on, a common transformer core. A third coil, the detector coil, is wound onto the transformer core and is connected directly to a sensitive relay (trip coil).

When the phase and neutral currents are balanced as they should be in a healthy circuit, they produce equal and opposing fluxes in the transformer core, and therefore resulting in no current in the detector coil.

If more current flows in the phase conductor than the neutral, an out of balance flux will be produced, thus inducing a current in the detector coil which will supply the trip coil, causing the relay to operate the tripping mechanism and open the main contacts.

Normally, the reason for more current flowing in the phase than the neutral conductor is because some current has returned to earth via an earth fault.

Residual Current Devices *(531-02)*

Where an RCD may be operated by other than a skilled or instructed person, it should be designed or installed so that adjustment of the setting or calibration of its rated trip current or time delay is not possible without the use of a key or tool. Visible indication of its setting or calibration is required by Regulation 531-02-10.

Test Button

The operation of an RCD by depressing the test button proves that the mechanical parts of the device are working correctly. The users of an installation are advised to carry out this test quarterly (see Regulation 514-12-02).

Selectivity

RCDs are completely selective in operation of the circuit they protect and are unaffected by parallel earth paths.

Application

When a device is fitted as protection against indirect contact but separately from overcurrent protection, the device must be capable of withstanding, without damage, any thermal and mechanical stress which may occur under short circuit conditions on the load side of the device.

If the disconnection times are to be achieved by the use of an RCD in an installation supplied from a TN or TT system, the product of the rated residual operating current (IΔn) and the earth fault loop impedance (Ω) must not exceed 50 volts.

When installing RCDs, Regulation 314-01-01 requires that circuits should be arranged to avoid danger and minimise inconvenience caused by operation of the device (see also Regulation 531-02-04 regarding unnecessary tripping).

Note: The use of an RCD is excluded for automatic disconnection when the system is TN-C. In such a case there is no difference between phase and neutral currents because there is no separate path for neutral and earth leakage currents.

EARTHING ARRANGEMENTS AND PROTECTIVE CONDUCTORS

Purpose of Earthing

The earth can be considered to be a large conductor that is at zero potential.

The purpose of earthing is to connect together all metalwork (other than that which is intended to carry current) to earth so that dangerous potential differences cannot exist either between different metal parts, or between metal parts and earth.

By connecting the non-current carrying metalwork to earth, a path is provided for leakage current that can be detected and, if necessary, interrupted by the following devices:

- fuses
- circuit breakers
- residual current devices (RCDs)

Types of Electrical System

An electrical system consists of a single source of electrical energy and an installation.

For certain purposes of the IEE Wiring Regulations, different types of electrical system can be identified depending on the relationship of both the source and the exposed conductive parts of the installation to earth. These systems are:

- **TN system**
  A system having one or more points of the source of energy directly earthed, the exposed conductive parts of the installation being connected to that point by protective conductors

- **TN-C system**
  In which neutral and protective functions are combined in a single conductor through the system

- **TN-S system**
  This has separate neutral and protective conductors throughout the system

- **TN-C-S system**
  In which neutral and protective functions are combined in a single conductor in part of the system

- **TT system**
  A system having one point of the source of energy directly earthed, the exposed conductive parts of the installation being connected to earth electrodes electrically independent of the earth electrodes of the source.

- **IT system**
  A system having no direct connection between live parts and earth, the exposed conductive parts of the electrical installation being earthed

The first letter allocated to each system indicates the **supply** earthing arrangements:

    T   one or more points of the supply are directly connected to earth

    I   supply system not earthed, or one point earthed through a fault limiting impedance

The second letter indicates the **installation** earthing arrangements:

    T   exposed conductive parts connected directly to earth.

    N   exposed conductive parts connected directly to the earthed point of the source of the electrical supply. (The point where neutral normally originates.)

The third and fourth letters, where appropriate, indicate the **relationship between the neutral and protective conductors**:

    S   separate neutral and protective conductors

    C   neutral and protective conductors combined in a single conductor

Further details can be found in the section entitled 'System Earthing Arrangements'.

Connections to Earth

The earthing arrangement of an installation must be such that:

- the value of impedance from the consumer's main earthing terminal to the earthed point of the supply for TN systems or to earth for TT and IT systems is in accordance with the protective and functional requirements of the installation and expected to remain continuously effective

- earth fault and protective conductor currents which may occur under fault conditions can be carried without danger, particularly from thermal, thermomechanical and electromechanical stresses

- they are robust or protected from mechanical damage appropriate to the assessed conditions

The installation should be so installed as to avoid risk of subsequent damage to any metal parts or structures through electrolysis.

Voltage Ranges

Extra Low Voltage (ELV)

0 V to 50 V AC or 0 V to 120 V ripple-free DC (whether between conductors or to earth).

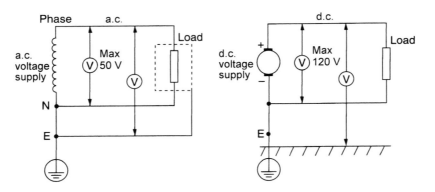

Extra low voltage systems

Low Voltage

Exceeding ELV, but not exceeding 1000 V AC or 1500 V DC between conductors or 600 V AC (RMS) or 900 V DC between conductors and earth.

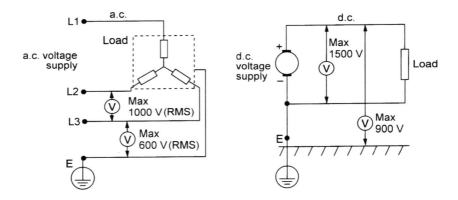

Low voltage systems

System Earthing Arrangements

TN-S Systems

This is likely to be the type of system used where the distributor's installation is fed from underground cables with metal sheaths and armour. In TN-S systems, the consumer's earthing terminal is connected by the distributor to the distributor's protective conductor (i.e. the metal sheath and armour of the underground cable network) which provides a continuous path back to the star point of the supply transformer, which is effectively connected to earth.

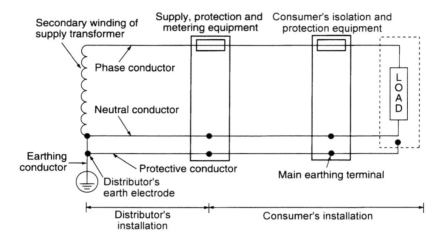

TT Systems

This is likely to be the installation used where the distributor's installation is fed from overhead cables. With such systems, no earth terminal is provided. An earth electrode for connecting the circuit protective conductors to earth has to be provided by the consumer. An effective earth connection can be difficult to obtain and a residual current device should be installed in addition to any overcurrent protective devices.

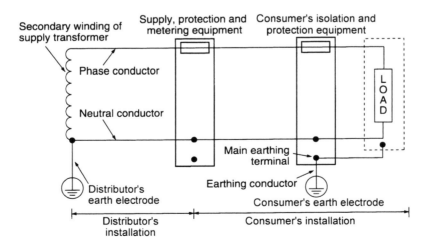

TN-C-S Systems

When the distributor's installation uses a combined protective and neutral (PEN) conductor, this is known as a TN-C supply system. Where consumer's installations consisting of separate neutral and earth (TN-S) are connected to the TN-C supply system, the combination is called a TN-C-S system. This is the system usually provided to the majority of new installations, referred to as a PME system by the distributor.

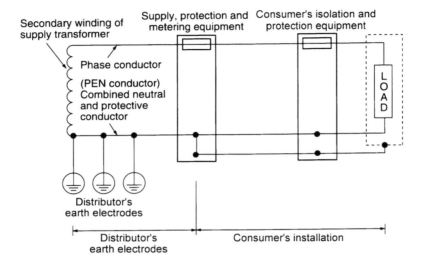

TN-C Systems

Where a combined neutral and earth conductor (PEN conductor) is used in both the supply system and the consumer's installation this would be referred to as a TN-C system.

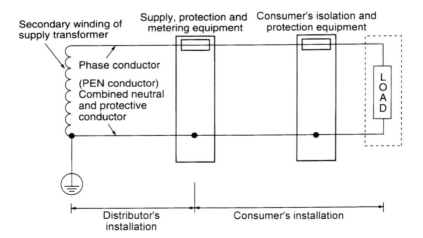

Regulation 8(4) of the Electricity Safety, Quality and Continuity Regulations 2002 prohibits a consumer from combining the neutral and protective functions in a single conductor within the consumer's installation.

Earthing Arrangements and Terminations

The diagrams below show what the three most common earthing arrangements look like in practice.

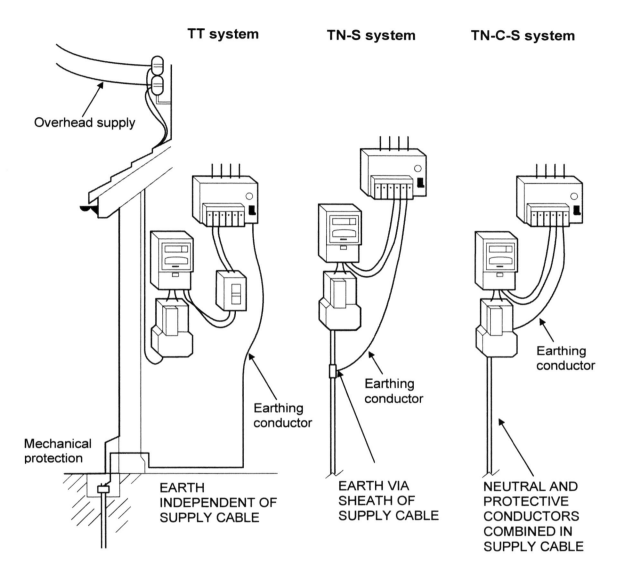

In the United Kingdom, distributors have to comply with the Electricity Safety, Quality and Continuity Regulations 2002.

Full discussions with the relevant distributor are essential when planning or installing a customer's installation in order to obtain the specifications for any special requirements, e.g. size of earthing conductor and main bonding conductors, values of earth loop impedance (Ze) and prospective fault current for the electrical supply.

Main Earthing Terminals or Bars

A main earthing terminal must be provided in every installation to enable the earthing conductor to connect to:

- circuit protective conductors
- main bonding conductors
- functional earthing conductors (if required)

Provision must be made for disconnection of the earthing conductor for test measurement of the earthing arrangements.

The method of disconnecting the earthing terminal from the means of earthing must be such that it can only be effected with the use of tools, and may conveniently be combined within the main earthing terminal.

Protective Conductors

Cross-sectional Areas

The minimum cross-sectional area (csa) of protective conductors can be

- Selected (as 543-01-04)

If the protective conductor does not form part of a cable the cross sectional area should not be less than:

- 2.5 mm² if sheathed, or otherwise provided with mechanical protection
- 4 mm² where mechanical protection is not provided

If the protective conductor csa is to be selected then Table 54G, BS 7671 is used.

Table 54G establishes the minimum csa of protective conductor in relation to the csa and material of the associated phase conductor. For example:

Cross-sectional area of phase conductor. SIZE (s) mm²	Minimum cross-sectional area of the corresponding protective conductor. SIZE(s) mm²
	If the protective conductor is of the same material as the phase conductor
$s \leq 16$	s
$16 < s \leq 35$	16
$s > 35$	$\dfrac{s}{2}$

Types of Protective Conductor

A protective conductor may be one or more of the following:

- a single core cable (colour GREEN and YELLOW)
- a conductor in a cable

The circuit protective conductors of ring final circuits should be installed in the form of a ring with both ends connected to the earth terminal at the origin of the circuit, e.g. distribution board or consumer unit (543-02-09).

Main Equipotential Bonding Conductors

Main bonding conductors should connect the following to the installation's main earthing terminal (MET):

- water service pipes
- fuel installation pipes (e.g. gas and oil)
- other service pipes and ducting
- central heating and air-conditioning systems

The use of plastic pipework for installations within buildings can affect equipotential bonding, therefore:

- If the incoming service pipes are plastic, and the pipes within the installation are also plastic, they do not require main bonding
- If the incoming service pipes are plastic, but the pipes within the installation are metal, the main equipotential bonding **must** be carried out to the metal installation pipes

The main bonding conductors should be not less than half the cross-sectional area of the earthing conductor; the minimum size being 6 mm² with a maximum size of 25 mm².

Where PME conditions apply, the cross-sectional area should be in accordance with Table 54H (see below), which gives the minimum cross-sectional area of main bonding conductors in relation to the neutral conductor of the incoming supply. The conductor sizes are for copper conductors or other conductors affording equivalent conductance.

Supply neutral conductor cross-sectional area	Main equipotential bonding conductor cross-sectional area
35 mm² or less	10 mm²
Over 35 mm² up to 50 mm²	16 mm²
Over 50 mm² up to 95 mm²	25 mm²
Over 95 mm² up to 150 mm²	35 mm²
Over 150 mm²	50 mm²

Note: The local distributor's network conditions may require larger conductors.

The extraneous conductive parts within an installation, such as gas, oil and water services, must be at the same potential as the exposed conductive parts, i.e. the metalwork of the electrical installation. This creates an earthed equipotential zone. This is achieved by connecting all the exposed conductive parts of the electrical installation to the main earthing terminal by the circuit protective conductors (cpcs) and by installing main bonding conductors from the main earthing terminal to the gas, oil, water and other services at their point of entry to the premises, as illustrated on the next page.

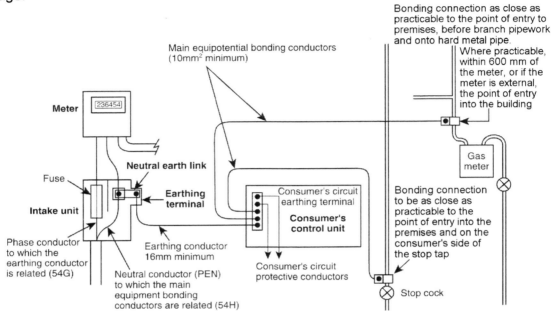

Main equipotential bonding to gas, oil, water and other services should be made as close as practicable to the point of entry of the service to the building, the bonding being applied on the consumer's side of any meter, stopcock or insulating section and before branch pipework. The connection is made to pipework using earth clamps to BS 951 that will not be affected by corrosion at the point of connection.

If there is a meter, the bonding connection should be on the consumer's side, preferably within 600 mm of the meter outlet before any branch pipework. If the meter is external to the building then, ideally, the connection should be made as close as practical to the point of entry of the service to the premises.

The main bonding conductors should be separate (as shown) or a single conductor may be used provided it passes **unbroken** through the connection at one service and directly on to the other.

The connection of a bonding conductor to metal pipework is usually by means of an earth clamp to BS 951. The clamp selected should be suitable for the environmental conditions at the point of connection. This is typically identified by the colour of the clamp body or a coloured stripe on the warning label, where:

- a RED stripe on the label indicates it is only suitable where the conditions are non-corrosive, clean and dry (as a guide, hot pipes only)

- a BLUE or GREEN stripe on the label indicates that the clamp is suitable for all conditions (including corrosive and humid)

The clamp must be adequate for the size of conductor being connected and it must have a label attached with the words:

'Safety electrical connection – do not remove'.

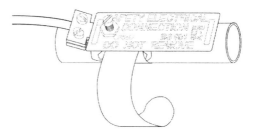

If a gas meter is fitted in an external, semi-concealed box and it is not practicable to provide a bonding connection at the point of entry due to the gas installation pipe entering the premises at low level (i.e. below the floorboards or buried in a concrete floor) then it is possible to make the bonding connection within the meter box itself.

The bonding conductor, when returning inside the premises, must pass through a separate hole above the damp course (with the hole sealed on both sides). The Gas Safety (Installation and Use) Regulations 1994 prohibit the bonding conductor passing through the same hole as a gas pipe.

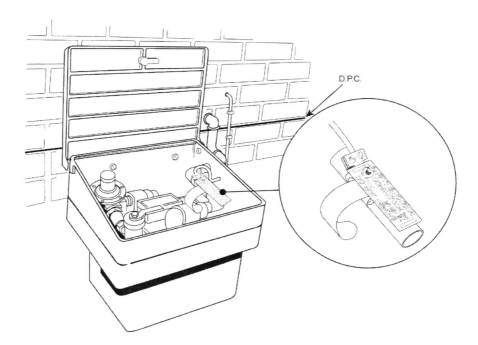

A connection made to the gas installation pipe externally between the meter box and entry point to the building would encourage the risk of corrosion and mechanical damage.

Where it is not practicable to provide a bonding connection, where the oil line is exposed at the point of entry, for example, where an existing oil supply pipe enters the premises below ground level or buried in a concrete floor, then it is possible to make the bonding connection immediately upon entry to the building.

In the case of oil fire valves BS 5410-1: 1997 allows for the location to be (only in these exceptional circumstances in existing installations) at the point where the oil line is first exposed internally.

Typical Earthing and Bonding Conductor Arrangements for Domestic Installations

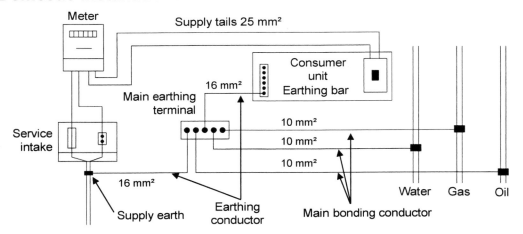

TN-S earthed to armour or metal sheath of the distributor's cable

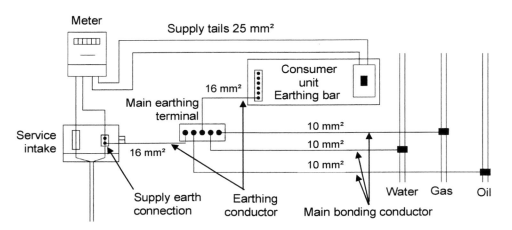

TN-C-S earthed using combined neutral and earth conductor of the distributor's cable

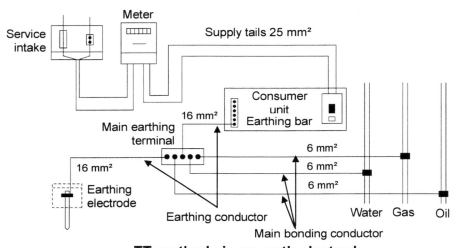

TT earthed via an earth electrode

Main bonding conductors can be installed separately or as an unbroken loop.

Note: The local distributor's network conditions may require larger conductors.

If there is a water meter on the consumer's side of, and close to, the stop tap then the main bonding conductor could be connected directly after the meter. Otherwise, the main bonding connection should be as close as possible to the point of entry on the consumer's side of the stop tap as normal.

If a water meter is installed within the premises on the consumer's side of, and some distance from, the stop tap then the water meter should be bridged by a bonding conductor. This is to prevent damage to the working parts of the meter in the event of a fault current flowing through that section of pipework in which the meter is connected.

Supplementary Bonding Conductors

If connecting exposed conductive parts, a supplementary bonding conductor (if sheathed or otherwise mechanically protected) must have a conductance of not less than the smallest protective conductor connected to the exposed conductive parts, subject to a minimum of:

- 4 mm² if mechanical protection is not provided

If connecting an exposed conductive part to an extraneous conductive part, a supplementary bonding conductor (if sheathed or otherwise mechanically protected) must have a conductance of at least half that of the protective conductor to the exposed conductive part, subject to a minimum of:

- 4 mm² if mechanical protection is not provided

In situations where a supplementary bonding conductor connects two extraneous conductive parts, neither of which are connected to an exposed conductive part, the minimum cross-sectional area of the supplementary bonding conductor must be:

- 2.5 mm² if sheathed or mechanically protected
- 4 mm² if mechanical protection is not provided

Supplementary bonding can be provided by conductors, or by permanent and reliable conductive parts or by a combination of the two.

Supplementary bonding conductors should connect together the exposed conductive parts of equipment of the relevant circuits (including the earth terminals of any socket outlets) and any extraneous conductive parts, and are required:

- where disconnection times cannot be met (although, as an alternative, RCDs may be installed)
- at locations of increased shock risk such as those covered by BS 7671-6, e.g. bath and shower rooms, swimming pools, etc.

Supplementary bonding where disconnection times cannot be met

The less common use of local supplementary bonding is where disconnection times cannot be met. Regulation 413-02-04 gives the option of:

- local supplementary bonding as 413-02-27 and 413-02-28, or
- protection by use of an RCD

Although supplementary bonding is accepted by Regulation 413-02-04, the normal choice would be the use of an RCD.

Where local supplementary bonding is used to comply with 413-02-04, it should connect together:

- the exposed conductive parts of equipment in the circuits concerned
- the earth terminal of any socket outlets or other accessories
- any extraneous conductive parts (e.g. gas pipe)

Supplementary bonding in areas of increased shock risk

In domestic premises, the areas identified as having increased shock risks are usually bath or shower rooms and around swimming pools.

Further measures would be required to support flexible wiring from above or the provision of additional mechanical protection. Providing RCD protection for circuits supplying portable equipment should also be considered.

Supplementary bonding in other locations

There is no specific requirement in BS 7671 to carry out supplementary bonding in areas like domestic kitchens, toilets and cloakrooms that may contain items like sinks, metal pipes and washbasins (BUT NOT A BATH OR SHOWER).

Supplementary bonding can be found at sinks and/or washbasins that are installed in areas other than those that contain a bath or shower. This could be due to previous editions of the IEE Regulations requiring sinks to be bonded where extraneous conductive parts were not reliably connected to the main bonding. Therefore, the bonding was applied whether required or not.

Note: Metal waste pipes in contact with earth should be main bonded to the main earthing terminal.

Plastic Pipework

The introduction of plastic pipework within buildings for hot and cold water and central heating systems has had an effect on supplementary bonding requirements. Supplementary bonding is not required to metal parts supplied by plastic pipes, e.g. hot and cold taps supplied by plastic pipework.

Supplementary bonding in bath and shower rooms

Local supplementary bonding should be used to connect the terminal of the protective conductor of circuits supplying Class I and Class II equipment in Zones 1, 2 or 3 and extraneous conductive parts in those zones, including:

- metal water, gas and other service pipes, metal waste pipes

- metal central heating pipes and air-conditioning systems

- accessible metal structural parts of the building (metal door and window frames etc. are not treated as extraneous conductive parts unless they are connected to structural metal parts of the building)

- metal baths and shower trays

An example of supplementary bonding in a bathroom with metal pipework is shown below.

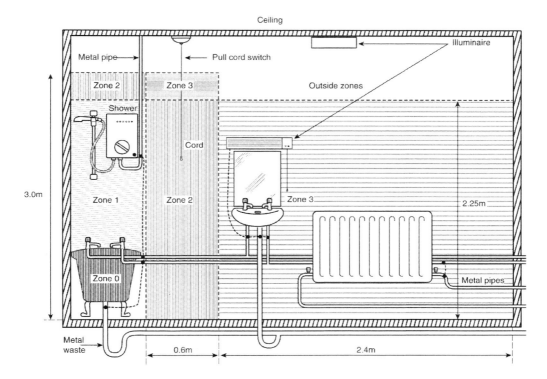

This bonding may be provided in close proximity to the location, e.g. an airing cupboard adjoining the bathroom.

The requirements for supplementary bonding apply to Zones 1 and 2 of a room other than a bath or shower room, e.g. bedroom where a shower cubicle is installed.

Regulation 547-03-04 permits supplementary bonding to be provided by:

- a conductor, or
- a conductive part of a permanent and reliable nature (e.g. metal pipework with soldered or compression type fittings that are electrically continuous), or
- a combination of these

If using metal pipework that has push fit type fittings (which do not provide electrical continuity) as a supplementary bonding conductor, the joints in the pipework will need to be bridged with a bonding conductor (see below).

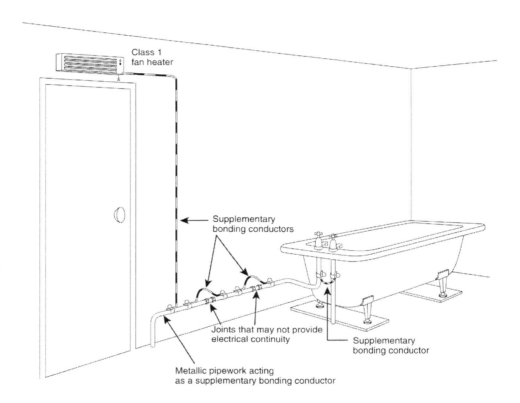

Plastic Pipework

The introduction of plastic pipework and push fit type fittings for use on metal pipe have had an effect on supplementary bonding requirements, particularly in bathrooms.

For plastic pipe installations, there is no need to supplementary bond metal fittings supplied by plastic pipe (e.g. hot and cold taps or radiators supplied by plastic pipes).

A metal bath **not** connected to any extraneous parts (e.g. structural steelwork and supplied by plastic hot and cold pipework and with a plastic waste pipe) does not require supplementary bonding (as shown on the next page).

The use of plastic pipe can create a safer electrical installation. Installing supplementary bonding to equipment connected to plastic pipework could reduce levels of electrical safety, not increase them.

An example of supplementary bonding in a bathroom with plastic pipework is shown below.

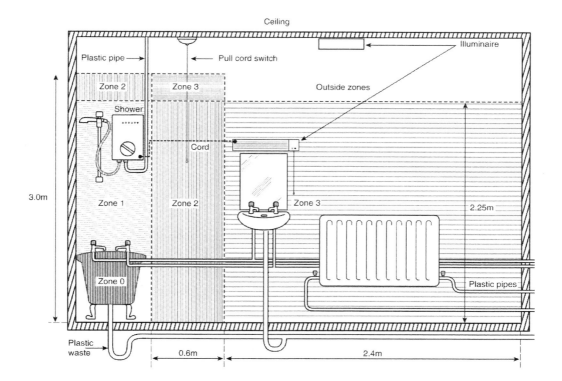

Earthing Clamps

Earth clamps to BS 951 are commonly used for primary equipotential bonding. Specially manufactured clamps are designed to make a secure and sound electrical contact with pipes carrying main services. The method to be adopted should be as follows:

- select a size of clamp to suit the size of pipe to be bonded

- clean the pipe using emery cloth or wire wool

- place a strap around pipe and through the clamp and tighten using a screwdriver; tighten down the locking nut (also securing the warning label)

- remove the insulation from the bonding conductor and connect to terminal on clamp (it is preferable to use a crimp terminal if possible to reduce the risk of the conductor being disconnected) and tighten; avoid cutting the bonding conductor when looping to another service pipe

- check that the earthing clamp and bonding conductor are tight

Termination to Earth Electrodes

The connection of earthing conductors to electrodes requires adequate insulation where they enter the ground, to avoid possible dangerous voltage gradient at the surface. All electrode connections must be thoroughly protected against corrosion and mechanical failure.

It is important that the electrode is made accessible for inspection purposes, and a label should be fitted at or near the point of connection.

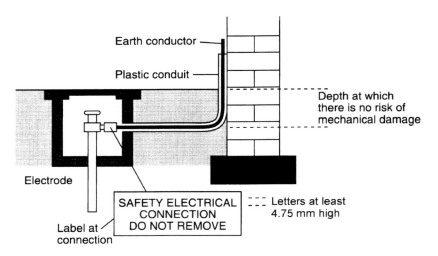

PVC WIRING, INSTALLATION TECHNIQUES AND CABLE SELECTION

PVC Cables

PVC (Polyvinyl chloride) insulated and sheathed cables are used extensively for lighting and heating installations in domestic dwellings, being generally the most economical method of wiring for this type of work.

Types of PVC Cable and Cord

Grades of PVC and their use in cords and cables

PVC compounds used for cords and cables are described in BS 6746C 1993. Several grades of compound are detailed in this standard for both insulation and sheathing requirements. PVC compounds are thermoplastic by nature and consequently are designed to operate within a prescribed temperature range. Grades of PVC can therefore be selected to suit particular environmental temperatures, the maximum conductor temperature for heat-resisting grade PVC being 85°C and the lowest operating temperature grade PVC below minus 30°C.

The majority of wiring installations, however, use a general purpose grade of PVC which has a maximum operating temperature of 70°C; this grade of PVC should not be installed or flexed when the air temperature is nearing 0°C.

Note: As from 31 March 2006, the cable core colours described below will change in line with the identification requirements of the European Electrical Standards Body, CENELEC. For more information on the new harmonised cable core colours, please see pages 111-113.

SINGLE-CORE PVC INSULATED UNSHEATHED CABLE

Application

Designed for drawing into trunking and conduit.

Construction
PVC insulated solid or stranded copper conductor, colours include:-
Old colours: red, yellow, blue, black, green &yellow
New colours:- brown, black, grey, blue, green & yellow
- other colours available

SINGLE-CORE PVC INSULATED AND SHEATHED CABLE

Application

Suitable for surface wiring where there is little risk of mechanical damage. Single-core is used for conduit and trunking runs where conditions are onerous.

Construction

PVC insulated and PVC sheathed solid or stranded plain copper conductor.

Old core colours:- black or red
New core colours:- brown or blue

Sheath colours: black, red, grey, white and other colours available.

SINGLE-CORE PVC INSULATED AND SHEATHED CABLES WITH CPC

Application

For domestic and general wiring where a circuit protective conductor is required for all circuits.

Construction

PVC insulated plain copper conductor laid parallel with an un-insulated plain copper circuit protective conductor, sheathed overall with PVC compound.

Old core colour: red.

New core colour: brown

CPC plain copper

Sheath colour: grey, white (old)
New: grey, white (white sheath on new thermosetting cable with low smoke (LSF) properties.

PVC INSULATED AND SHEATHED FLAT WIRING CABLES

Application

For domestic and industrial wiring. Suitable for surface wiring where there is little risk of mechanical damage.

Construction

Two and three-core cables with the inclusion of an un-insulated plain copper circuit protective conductor between the cores of twin cables and between the yellow and blue cores of three-core cables.

Core colours:

Two core + earth
Old – red and black
New – brown and blue

Three core + earth
Old – red, yellow and blue
New – brown, black and grey

All CPC Plain Copper

Sheath colours: grey and white (old)
New – grey and white (white sheath on new thermosetting cable with low smoke (LSF) properties.

PVC INSULATED AND SHEATHED FLEXIBLE CORDS

Application

General purpose indoors or outdoors in dry or wet locations e.g. portable tools, washing machines, vacuum cleaners, lawnmowers. Should not be used where the sheath can come into contact with hot surfaces. Not suitable for temperatures below 0°C.

Multi-core versions of this cable up to 20 cores have uses in control equipment.

Construction

PVC insulated plain copper flexible conductors laid up and PVC sheathed.

Core colours: brown phase, blue neutral and green/yellow earth, for core combinations up to five cores. Above this, cores are numbered with black numerals on white core insulation.

Sheath colours: black, white or orange.

RIBLITE' PVC INSULATED AND SHEATHED FLEXIBLE CORDS

Application

General purpose indoors or outdoors in dry or wet locations e.g. portable tools, washing machines, vacuum cleaners, lawnmowers. Should not be used where the sheath can come into contact with hot surfaces. Not suitable for temperatures below 0°C.

Construction

PVC insulated plain copper flexible conductors laid up with textile wormings and sheathed with PVC compound, with a ribbed finish.

Core colours: twin, brown and blue; three-core, brown, blue and green/yellow.

Sheath colours: grey, white or blue.

HEAT-RESISTING PVC INSULATED AND SHEATHED FLEXIBLE CORDS

Application

Suitable for use in ambient temperatures up to 85°C, e.g. for use with heating appliances etc.

Construction

Plain copper flexible conductors insulated with heat resisting (HR) PVC and HR PVC sheathed.

Core colours: single-core, brown or blue; twin, brown and blue; three-core, brown, blue and green/yellow; four core, brown, brown and green/yellow.

Sheath colour: white.

PVC INSULATED AND SHEATHED FLAT TWIN FLEXIBLE CORDS

Application

Intended for light duty indoors, for Class II double insulated equipment such as table lamps, radios and TV sets where the cable may lie on the floor. Not for use with heating appliances.

Construction

Plain copper flexible conductors PVC insulated, two cores laid parallel and sheathed overall with PVC.

Core colours: brown and blue.

Sheath colour: white, black etc.

PVC INSULATED AND SHEATHED HEAT RESISTING (HR) LIGHT DUTY FLEXIBLE CORDS

Application

Sometimes known as pendant flexibles, these are used for lighting.

Construction

Two or three-core cords having plain copper flexible conductors PVC insulated and sheathed.

Core colours: twin, brown and blue; three-core, brown, blue and green/yellow.

Sheath colour: white.

PVC INSULATED BELL WIRE

Application

For wiring to bells, alarms and other indicators which operate at extra low voltage.

Construction

Single-core: one 0.85 mm plain soft copper conductor, insulated with 0.5 mm radial thickness PVC. Insulation colours: red, black, white, green or brown.

Twin-core: two 0.85 mm diameter plain copper wires are laid parallel and insulated overall with PVC compound to form a figure 8 section. The standard colour for this wire is white. Core identification is by a rib or coloured stripe formed on one core

HARMONISED CABLE CORE COLOURS

This information also appears in the IEE On-Site Guide, Appendix 11: Identification of Conductors.

Introduction

The requirements of BS 7671 have been harmonised with the requirements of the European Electrical Standards Body, CENELEC, regarding the identification of conductors and the identification of cores in cables and flexible cords. (See BS 7671, Tables 51 and 7A to 7E or the IEE On-Site Guide, Tables 11A to 11F).

These standards specify the cable core marking, including the cable core colours, for the CENELEC countries.

For single-phase installations:

> The RED phase and BLACK neutral
>
> are replaced by
>
> BROWN phase and BLUE neutral
>
> The protective conductor remains GREEN and YELLOW

Therefore, fixed wiring for single-phase installations will now adopt the same colours that three-core flexible cables and cords have used for many years.

Notes: Installations beginning after 31 March 2004 may use the existing or the new harmonised core colours, but not both. Installations starting after 31 March 2006 must use only the new harmonised colours.

Wherever an interface (connection) occurs between old and new cable core colours, a warning notice as 514-04-01 should be displayed, on or adjacent to the consumer unit or distributor board.

Alterations or Additions to Existing Installations

Single-phase installations

Alterations or additions to a single-phase installation do not need marking at the interface where old wiring is connected to new, providing that:

- the old cables are coloured RED for phase and BLACK for neutral, AND
- the new cables are coloured BROWN for phase and BLUE for neutral

Two-phase or three-phase installations

At a wiring interface between old core colours and new core colours, clear identification is required as follows:

Old and new conductors

Neutral conductors = **N**

Phase conductors = **L1, L2, L3**

BS 7671 Table 7A gives examples of conductor marking at an interface for additions and alterations to an AC installation identified with the old cable colours (see below).

Function	Old conductor colour	Old/new marking	New conductor colour
Phase 1	RED	L1	BROWN
Phase 2	YELLOW	L2	BLACK
Phase 3	BLUE	L3	GREY
Neutral	BLACK	N	BLUE
Protective conductor	GREEN & YELLOW		GREEN & YELLOW

For a three-phase installation, as an alternative to the BROWN, BLACK and GREY identification of the phase conductors shown above, three BROWN or three BLACK or three GREY conductors may be used. However, they must be marked L1, L2 and L3 or oversleeved BROWN, BLACK and GREY at their terminations.

Switch Wires

New installations or modification to existing installations

Where a PVC-insulated and sheathed (twin and earth) cable is used as a switch wire and the cores, both being used as phase conductors, are coloured BROWN and BLUE:

- the BLUE conductor must be oversleeved BROWN or marked L at its terminations.

- the bare cpc must be oversleeved GREEN and YELLOW as normal

Intermediate and Two-Way Switch Wires

New installations or modification to existing installations

Where a three-core and earth cable with core colours BROWN, BLACK and GREY is used, and all three conductors are used as phase conductors:

- The BLACK and GREY conductors should be oversleeved BROWN or marked L at their terminations.

- The bare cpc being oversleeved GREEN and YELLOW as normal.

Phase Conductors

New installations or modification to existing installations

The colours and markings of phase conductors should be as BS 7671 Table 51.

For control circuits, extra low voltage and other applications:

- The phase conductor could be coloured BROWN, BLACK, RED, ORANGE, YELLOW, VIOLET, GREY, WHITE, PINK or TURQUOISE and marked L.

The neutral or mid-wire should be coloured BLUE and marked N.

Installing PVC Cables

Cables are fixed at intervals using plastic clips that incorporate a masonry type nail. The maximum spacing of clips for cables run on the surface is not specified by the IEE Regulations. Regulation 522-08-04 does state, however, that a conductor or cable shall be properly supported at intervals so that it does not suffer damage caused by its own weight. When bending PVC cable around corners, the radius of the bend should be such that the cable or conductor does not suffer damage (Regulation 522-08-03).

Guidance regarding the spacing of cable supports and the minimum internal radii of bends can be found in Appendix 4 of the IEE On-Site Guide.

For spacing of clips for cables in accessible positions refer to Table 4a.

Example

Overall diameter of cable	Horizontal	Vertical
Not exceeding 9 mm	250 mm	400 mm

The minimum internal radii of beads refer to Table 4e.

Example

Thermoplastic PVC insulated	Non armoured	Overall diameter	Factor
Flat construction	Yes	Not exceeding 10 mm	3

Clipping PVC Cables

1. In order to ensure a neat appearance PVC cable should be pressed flat against the surface between cable clips.

2. The cable should be formed by running the thumb against the surface of the cable, as illustrated.

3. Another method of forming the cable is to run the palm of the hand along the surface of the cable as illustrated.

4. This sequence of forming the cable should be carried out after inserting the last cable clip and before fixing the next cable clip.

5. When a PVC cable is to be taken round a corner or changes direction, the bend should be formed using the thumb and fingers as shown.

6. Care must be taken to ensure that the bend does not cause damage to the cable or conductors and that cable clips are spaced at appropriate intervals. See Regulation 522-08.

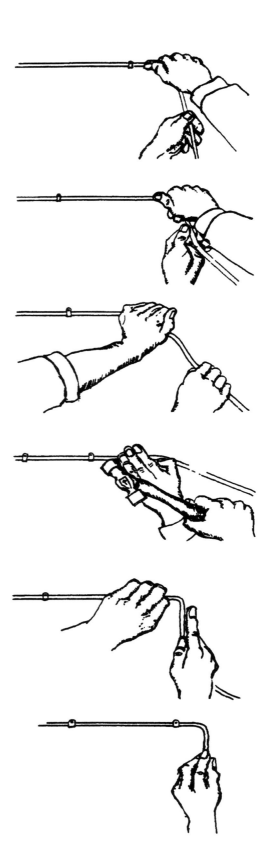

Surface Wiring

Where PVC cables are on the surface, the cable should be run directly into the electrical accessory, ensuring that the outer sheathing of the cable is taken inside the accessory.

Cable Runs

Cable runs should be planned so as to avoid cables having to cross one another, which would result in an unsightly and unprofessional finish.

Concealed Wiring

If the cable is concealed, a flush box is usually provided at each control or outlet position.

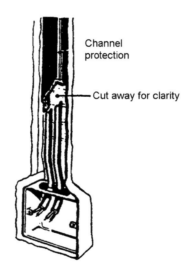

Embedded PVC Cables

In instances where wiring is installed during the course of the construction of the building, oval conduits or metal or plastic channelling should be used to protect cables from damage.

Where walls are already plastered a chase must be cut in the plaster and brickwork. Sheathed cables should be securely fixed and protected by channelling. Plaster mixes are used to make good.

Marking Out Chases

When chasing into building surfaces the aim should be to minimise disturbance to the surface. This is best achieved by marking off parallel lines just wide enough for the job in hand.

Safety Precautions

Safety goggles must be used whenever a chase is cut to prevent flying particles causing injury to the eyes. Suitable gloves should also be worn.

A plastic sheet should be placed on the floor to catch most of the debris and to make the job of clearing up easier.

Chase Work Using Hand Tools

Cutting a Chase in Plaster and Brick

1. Mark and chase by drawing parallel lines to show the width of the chase to be cut and the size of terminal box to be fitted.

   Lay a plastic sheet on the floor to catch the bulk of the debris.

   Put on safety goggles and use a pair of gloves.

   Use a wide flat chisel and light taps from the hammer to outline the chase and terminal box.

   Angle the chisel slightly inwards.

2. Chisel through the plaster down to the brickwork and lever off the plaster from the area being chased to expose the brickwork underneath.

3. Check the thickness of the cables and other mechanical protection (channelling, clips, over conduit, etc.) to be laid against the plaster.

   The level of the cables or channelling should be at least 5 mm below the surface of the plaster for proper concealment.

Channel protection

5 mm min. cover

4. Where chasing into the brickwork is necessary, use the two-speed electric drill at the slowest speed and a masonry drill (No. 10 or 12). Drill holes right around the area cleared of plaster.

   To obtain the correct depth of drilled hole use a block of wood of the required thickness as a gauge.

   Simply drill centrally through the block of wood taking care to hold it firmly when drilling. In this way holes of the correct depth can be drilled each time. If a proper gauge can be fitted to the drill, use it in preference to a block of wood.

   Holes should not be more than 5 mm apart around the sides of the chase.

5. A third vertical row of holes down the centre of the chase may also be necessary to facilitate chiselling away the intervening brickwork.

   The area to be recessed for the terminal box will need to be honeycombed with drill holes, using a spacer block of reduced thickness to obtain holes of the correct depth for the box to be fitted.

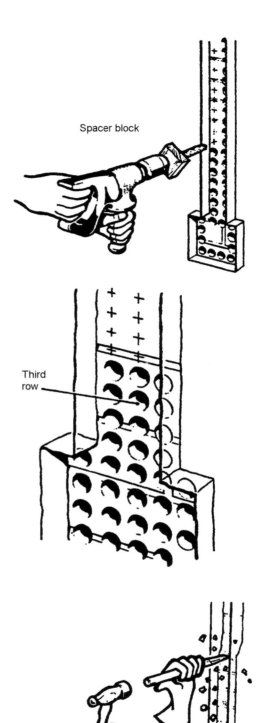

6. Use a sharp cold chisel and short sharp strokes from the hammer to even up the bottom and sides of the chase down to the required depth.

   Position yourself to one side of the chase when chiselling to keep out of the way of the flying particles and dust, but make certain that you can clearly see what you are doing.

7. Chisel out the recess for the terminal block again using the cold chisel.

   Complete the recess by chamfering away the vertical chase to allow for a smooth entry of the cables into the terminal box.

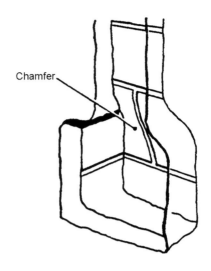

8. Check that the lip of the box is 5 mm below the level of the plaster surface.

   Position the terminal box so that it is level and plumb. Use a small spirit level to check.

   Mark out the drill holes, drill, plug and screw the terminal box into position.

9. Draw in the sheathed cables through the rubber grommet and fix channelling over the cables using galvanised or non-ferrous nails.

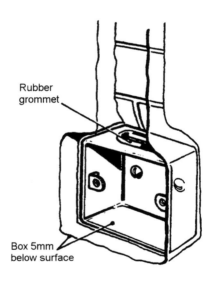

10. Mix up sufficient plaster and water to fill one chase at a time.

    Do the mixing on a board adding the water (1 part by volume to 6–7 parts of plaster) slowly to a hollow made in the centre of the heap of plaster. The mix should be quite dry but smooth (not lumpy).

11. Push some crumpled newspaper into the switch box to keep it clear of plaster. Wet the chase to prevent the plaster drying out too quickly and cracking.

    Place some of the plaster on a smaller board, hold it against the wall and starting at the bottom use the trowel to push small quantities of plaster at a time into the chase to provide a rough rendering coat to fill two-thirds of the space.

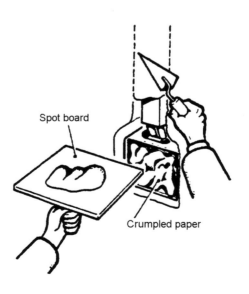

12. Next, use the tip of the trowel to fill and finish off the plastering around the terminal box and complete filling in the chase with plaster. For deep chases (in excess of 25 mm), it is advisable to wait 24 hours before completing the filling operation to allow the rendering coat to set until reasonably solid.

13. Sprinkle the plaster with minute quantities of water (using a brush dipped in clean water) and smooth out the surface of the plaster.

    Remove the paper from the accessory box and connect the switch, socket outlet or accessory

<blockquote>A</blockquote>

Depth of Chase

The Building Regulations require chases to positioned as not to impair the stability of the wall, especially where hollow block are encountered. The depth of chases should not be deeper than indicated in the table below.

Type of chase	Max depth of chase
Vertical	$\frac{1}{3}$ thickness of wall or leaf
Horizontal	$\frac{1}{6}$ thickness of wall or leaf

Note: The leaf refers to the material either on the outside or inside of a wall with a cavity.

Precautions when cables pass through walls, floors and ceilings

When cables are run under wooden floors or above ceilings, the cables should be fixed to the side of the joists. When run across the joists, the joists should be drilled sufficiently low that no damage could occur to the cable by screws and nails penetrating floorboards. The cable is run through these holes in the joists.

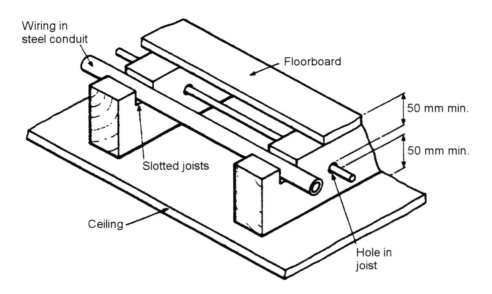

Support and protection for cables run under floorboards

A Notches and Holes

For a single family house up to three storeys high the requirements for notches and holes in simply supported floors and roof joist should be:

- maximum depth of notch should be 0.125 x joist depth
- not closer to the support than 0.07 of the span
- not further away from the support than 0.25 times the span
- maximum diameter of hole should be 0.25 x joist depth
- made at the neutral axis (centre)
- holes in the same joist should not be less than 3 diameters apart
- holes on centre line in zone between 0.25 and 0.4 x span

Note: No holes or notches are to be cut in any roof rafters, in order to prevent damage to the roof structure.

Keeping cables away from other service installations

Care must be taken when installing PVC cables to ensure that they are not allowed to come into contact with gas pipes, water pipes and any other non-earthed metalwork.

Terminating Cables and Flexible Cords

The entry of a cable end into an accessory is known as a termination. In the case of a stranded conductor, the strands should be twisted together with pliers before terminating.

Where possible, any single conductors should be folded to ensure an effective connection. Care must be taken not to damage the conductors.

The IEE Regulations require that a cable termination of any kind should securely anchor all the strands of the conductor and not impose any appreciable mechanical stress on the terminal or socket or any undue strain on the conductor itself.

A termination under mechanical stress is liable to loosen or disconnect. When current is flowing a certain amount of heat is developed, and the consequent expansion and contraction may be sufficient to allow a conductor under stress, particularly one under tension, to loosen or be pulled out of the terminal or socket.

Unless equipment manufacturers' instructions state otherwise, any conductors should preferably always be of sufficient length to allow them to be terminated at least one more time.

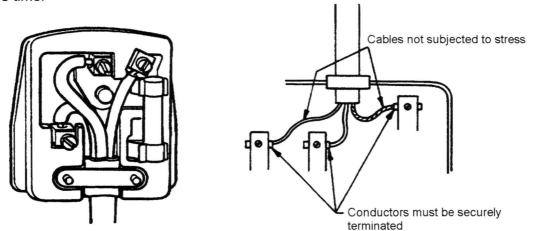

One or more strands, or wires, left out of the terminal or socket, will reduce the effective cross-sectional area of the conductor at that point. This may result in possible overheating due to introducing further resistance into the circuit. The same effect could occur with a loose connection.

Terminating flexible cords or cables

Just the minimum amount of insulation should be removed to achieve an effective connection, with the terminal screw firmly clamping the conductor. A good, clean, tight termination is essential.

The removal of insulation is preferably carried out with proprietary stripping tools for a cleaner, neater job, avoiding damage to the conductor, its insulation and sheath.

Method

1. Using side cutting pliers or a knife, slit the outer sheath from the end.

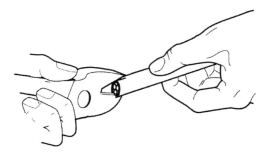

2. Peel back the two halves of insulated sheath for a suitable distance.

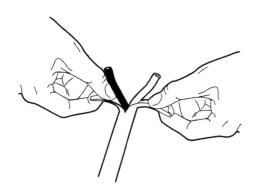

3. Cut off the sheath, neatly.

4. Examine the conductor insulation for damage.

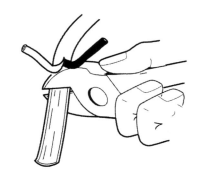

Cable-stripping Tools

1. Adjust screw to suit the diameter of the conductor. Push the end of wire into the tool so that the Vee slots close and cut through the insulation.
2. Remove severed insulation.
3. Examine the conductor for damage.

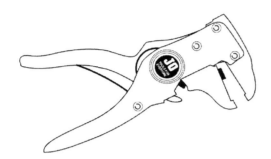 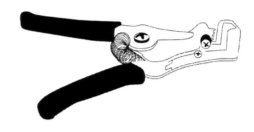

Cable-stripping tools

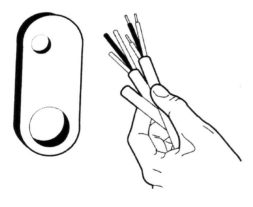 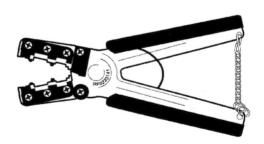

Flexible cable/wire stripper and cutter **Twin and earth**

Types of Terminal

There is a wide variety of conductor terminations. Typical methods of securing conductors in accessories are pillar terminals, screwheads, and nuts and washers. Push-in connectors are also increasingly common.

Pillar Terminals

A pillar terminal has a hole through its side into which the conductor is inserted and then secured by a set screw. If the conductor is small in relation to the hole, it should be doubled back. Care should be taken not to damage the conductor by excessive tightening.

Screwhead and Nut and Washer Terminals

When fastening conductors under screwheads or nuts, it is best to form the conductor end into an eye, using round-nosed pliers. The eye should be slightly larger than the screw shank, but smaller than the outside diameter of the screwhead, nut or washers. The eye should be placed in such a way that rotation of the screwhead or nut tends to close the joint in the eye. If the eye is put the opposite way round, the motion of the screw or nut will tend to untwist the eye, and will probably result in imperfect contact.

Strip Connectors

The conductors to be terminated are clamped by means of grub screws in connectors which are usually made of brass and mounted in a moulded insulated or porcelain block.

Just the minimum amount of insulation should be removed to achieve an effective connection so that the terminal screw firmly clamps the conductor. A good, clean, tight termination is essential in order to avoid a high resistance connection resulting in overheating of the joint.

Where possible, single conductors should always be folded.

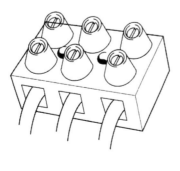 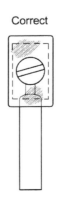 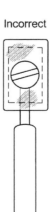

Correct   Incorrect

Terminating a Plug to Flexible Cord

Method

1. Remove the plug top.

2. Bring the flexible cord alongside the plug to measure the required amount of sheathing to be removed (or use manufacturer's stripping guide).

3. Cut the conductors to correct length.

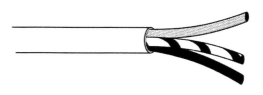

4. Remove sufficient insulation to expose conductor of correct length to ensure correct termination

5. Make 'L' (Brown), 'N' (Blue) and 'E' (Green/Yellow) connections, ensuring that the conductors go to the correct terminals.

    Just the minimum amount of insulation should be removed to achieve an effective connection and every strand or wire should be securely connected. Ensure that the terminal screws are tight.

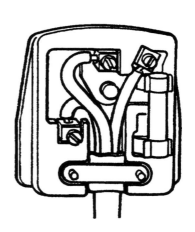

6. Clamp the flexible cord securely in or with the cord grip.

7. Replace the plug top.

Terminating Cable Ends to Crimp Terminals

In order to terminate conductors effectively crimp terminals are extensively used. This type of connection is often used in the termination of bonding conductors to earth clamps.

The terminals are usually made of tinned sheet copper with silver-brazed seams. The colour-coded crimp terminals represent the cable sizes they are designed for use with and they are typically:

- RED = 0.75 mm² – 1.5 mm²
- BLUE = 1.5 mm² – 2.5 mm²
- YELLOW = 4 mm² – 6 mm²

Crimp terminals

The heavy-duty crimping tool is made with special steel jaws which are adjustable in order that a range of cables and terminals can be crimped.

The ratchet crimping tool must be fully closed before the jaws will open to release the crimp terminal. This is to ensure correct connection.

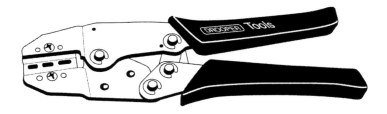

Ratchet crimp tool

Heavy duty crimp tool

Method

1. Remove the correct amount of cable insulation.

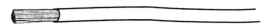

2. Place into the terminal.

3. Crimp using a crimping tool in accordance with the manufacturer's instructions.

4. Check the connection for soundness by holding the cable firmly and giving the terminal a sharp tug between the thumb and forefinger.

CABLE SELECTION

Conventional Final Circuit Design (using Table 7.1 of the IEE On-Site Guide)

Circuit design normally involves extensive use of tables and calculation that can get quite involved. The IEE On-Site Guide contains Table 7.1 for conventional circuits that enables the maximum cable run to be established to comply with the IEE Regulations without calculation.

In order to use the table safely, the proposed circuit should comply with the following assumptions.

The installation is supplied by one of the following systems:

- TN-C-S with a maximum loop impedance Ze of 0.35 Ω
- TN-S with a maximum loop impedance Ze of 0.8 Ω
- TT with RCDs installed as Section 3.6 of the IEE On-Site Guide

The final circuit is to be connected to a consumer unit at the origin of the installation.

The installation method used complies with those specified in the 16th Edition IEE Regulations Table 4A for:

- Method 1 sheathed cables clipped or embedded in plaster
- Method 3 cables run in conduit or trunking
- Method 6 PVC insulated and sheathed (twin and earth) cables in an insulating wall or above a thermally-insulating ceiling or single-core PVC insulated cables in conduit in a thermally-insulating wall.

Throughout the length of the circuit the temperature does not exceed 30°C.

Any grouping of cables will have to be taken into consideration (see Section 7, IEE On-Site Guide).

Protective device characteristics are in accordance with Appendix 3 of BS 7671 with a fault current tripping time for circuit breakers of 0.1s or less.

Using Table 7.1 – Example

A 13 A immersion heater circuit protected by a Type B 16 A circuit breaker wired in 2.5 mm² PVC sheathed cable with a 1.5 mm² cpc is supplied from a 230 V single-phase TN-C-S system. The maximum length of run is 33 m.

Radial Circuit

Current rating Amps	Cable CPC size mm²		Protective device	Cable installation method		Max. length in metres			
				PVC cable	Thermo setting cable	TN-S 0.4 s	TN-S 5 s	TN-C-S 0.4 s	TNC-S 5 s
16	2.5	1.5	Type B circuit breaker	M6	M6	33	33	33	33

Note: M6 indicates that the installation methods M1, M3 or M6 from Table 4A of BS 7671 may be used.

Circuit Design Example

An 8 kW electric shower is to be installed in the bathroom of a domestic dwelling, the length of proposed cable route is 30 m clipped to the surface supplied from the spare way in a consumer's unit fitted with Type B circuit breakers. The electricity supply is a 230 V TN-C-S system.

Design a suitable installation to comply with the requirements of the IEE Regulations using Table 7.1 of the IEE On-Site Guide to minimise the number of calculations required.

Step 1: Determine the full load current (I):

$$I = \frac{watts}{volts} = \frac{8,000}{230} = \textbf{34.8 A}$$

Step 2: Determine the design current.

No diversity allowance for the first two showers installed.

Design current I_b = **34.8 A**

Step 3: Select the type and size of protective device.

Since the consumer unit is fitted with Type B circuit breaker, use **Type B 40 A** since this is the next size of circuit breaker > 32 A.

Step 4: Apply correction factors, e.g. for temperature, etc.

None apply.

Step 5: Check circuit complies with shock protection requirements (although the regulations no longer give exclusive disconnection times for bathrooms and normal regulations apply, it would be prudent to continue to use 0.4 seconds due to the hazardous nature of the location).

Using Table 7.1 IEE On-Site Guide:

For a 10 mm² cable with a 4 mm² cpc using a 40 A Type B circuit breaker on a TN-C-S system.

The maximum length of run is **53 m**.

Since the actual cable run is 30 m, which is less than 53 m maximum, the circuit will comply with the IEE Regulations.

Summary

Before installing any conventional final circuit, the following must be considered:

- Can the consumer unit or distribution board carry the planned additional load?
- What type of protective device is to be used?
- What type of cable and installation method is to be used?
- What are the ratings of the protective devices available?
- What type of earthing arrangement is being used?
- What is the maximum disconnection time for the circuit; 0.4 s or 5 s?
- What isolation and switching requirements are necessary?
- What labels are required to be fitted?
- Is the earth loop impedance value below the values of 7.1(i) or 7.2.4(ii) of the On-Site Guide?
- Is an RCD or RCBO required? (Socket outlets for equipment outdoors or if the supply system is TT, a 30 mA RCD must be installed. Certain equipment in bath and shower rooms requires RCD protection.) All RCDs or RCBOs to comply with BS 4293, BS 7288, BS EN 61008 or BS EN 61009.

Note: On all systems where socket outlets supply equipment outdoors, protection by an RCD with a maximum operating current (IΔn) of 30 mA is required.

ARMOURED CABLES

PVC-SWA

PVC-SWA cables have been used extensively for supplies to and for wiring circuits to electric gates. The cable consists of single or multi-core PVC insulated conductors made of copper or aluminium with a PVC sheathed steel wire armour and a PVC sheath overall. This cable is manufactured to BS 6346: 1997.

SWA-XLPE

XLPE cables to BS 5467 are now used extensively in new installations. The cable has an outer sheath of PVC over the galvanized mild steel armour wires that have a PVC bedding over the XLPE insulated conductors.

XLPE has better insulation properties compared to PVC and this enables XLPE cables to have smaller external diameters for a given voltage rating.

XLPE cables can be used at up to 90°C maximum operating temperature (but must be limited to 70°C when installed in conduit, etc.).

Advantages:

- More pliable, lighter and easier to handle than paper-insulated, lead-covered cables.
- Termination of this cable is also relatively easy.

Disadvantages:

- This type of cable is more likely to sustain damage when installed where the temperature is below 1°C, or in conditions where the insulation is likely to split, or in the case of PVC-SWA, installed where the cable will be subjected to temperatures in excess of 70°C for long periods.

Sizes

Typical sizes range from 50 mm² to 100 mm² for single-core types and up to 400 mm² for two-, three- and four-core types. Multi-core armoured cables are available with the individual cores having white insulation but each core is individually numbered.

LSOH

Armoured cables with reduced flame propagation, low smoke and corrosive gas emission insulation are widely available. They are ideal for emergency electrical services, e.g. sprinkler pumps and smoke extractors, etc.

For further details consult manufacturers' data.

Installation

The cable can be laid directly in the ground, in ducts or fixed directly onto walls using cable cleats.

Terminating PVC-SWA cable

Terminations are made by stripping back the PVC sheathing and steel wire armourings, and fitting a compression gland which is then screwed into switchgear or control gear. An earthing tag (bonding ring) provides earth continuity between the armour of the cable and the box or panel.

TERMINATING PVC-SHEATHED SWA CABLE

1. Decide the position of the cable gland to expose the required conductor length for termination and to locate the point where the steel armour wires will require cutting.

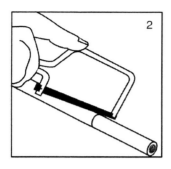
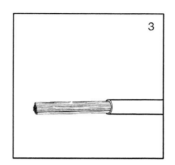
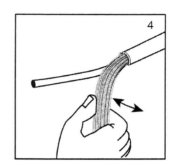
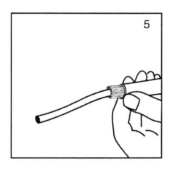
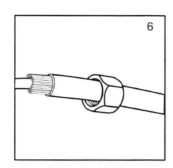
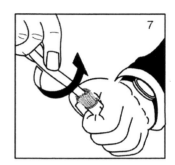
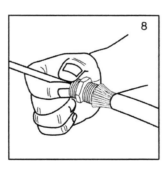

2. Using a hacksaw, work around the cable at the required position, cutting through the outer sheath and halfway through all the steel armour wires.

3. Remove the outer sheath.

4. Break off the armour wires by bending them back and forth (if you have not cut through the wire sufficiently it will tend to bend and not break off cleanly).

5. Using a knife, cut back a further small length of outer sheath to expose sufficient length of steel armour for an effective connection at the gland.

6. Place gland nut onto cable.

7. Move inner core in a circular outward motion to fan out armour, as shown.

8. Place main body of gland on cable, making sure that it passes inside the armour wires.

Continued

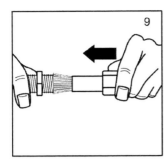 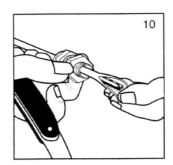 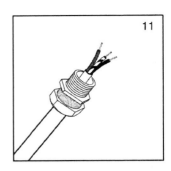

9. Move the gland nut up over the armour wires which are gripped between the gland body and the gland nut. Tighten the gland nut.
10. Remove the inner PVC sheath to expose the conductors. **N.B.** This inner sheath should extend for at least 3 mm beyond the end of the gland to protect the conductor insulation.
11. Remove sufficient insulation for each of the conductors to be correctly terminated.

TYPICAL EXAMPLES OF PVC/SWA CABLE

Design and construction

THREE-CORE CABLE, SINGLE WIRE ARMOURED and PVC OVERSHEATHED

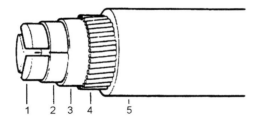

1. Solid aluminium conductor
2. PVC insulation
3. Taped bedding
4. Galvanised steel wire armour
5. PVC oversheath

FOUR-CORE CABLE, SINGLE WIRE ARMOURED and PVC OVERSHEATHED

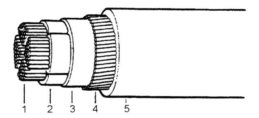

1. Shaped stranded copper conductor
2. PVC insulation
3. Extruded bedding
4. Galvanised steel wire armour
5. PVC oversheath

FOUR-CORE CABLE, ALUMINIUM STRIP ARMOURED and PVC SHEATHED

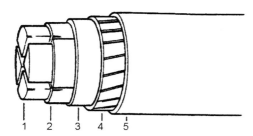

Conductors of shaped, solid aluminium are insulated with PVC. The cores are laid up and bound with PVC tapes followed by a single layer of aluminium strips and an overall PVC sheath.

1. Solid aluminium conductor
2. PVC insulation
3. Taped bedding
4. Aluminium strip armour
5. PVC oversheath

A1/A2 GLAND with OUTER SEAL

General purpose gland for plastic or rubber insulated, unarmoured plastic or rubber oversheathed cables.

Application

For general use indoors and outdoors.

Specification

Gland consists of: gland body (4), polychloroprene (PCP) seal (3), skid washer (2) and gland nut (1).

The cable oversheath is sealed by tightening the gland nut; the risk of torque being induced into the cable is eliminated by using a skid washer.

All metal parts are made of brass, accurately machined for ease of assembly.

SINGLE-CORE SECTORAL CABLE

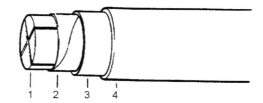

The conductor consists of four sector shaped solid aluminium conductors laid up together and bound. Extruded PVC insulation is applied over the conductor followed by a PVC oversheath. A similar armoured version is also available.

1. Solid aluminium conductor
2. Conductor binder
3. Extruded PVC insulation
4. PVC oversheath

Fitting sequence

(a) Screw item 4 into apparatus (secure with backnut if plain hole entry).
(b) Pass cable end through items 2 and 3 and then through 4.
(c) Engage items 2 and 4 and tighten up.

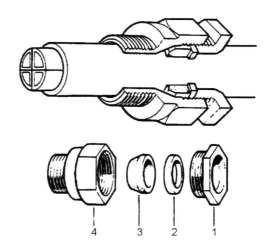

BW GLAND

Indoor gland for plastic or rubber insulated, single wire armoured, plastic or rubber oversheathed cables.

Application

For dry indoor situations.

Specification

Gland consists of: gland body with knurled integral armour clamping cone (1), and gland nut (2).

All parts are made of brass, accurately machined for ease of assembly.

Fitting sequence

(a) Screw item 1 into apparatus (secure with backnut if plain hole entry)

(b) Pass item 2 over cable before commencing to strip oversheath

(c) Remove oversheath to expose armour

(d) Pass cable end through item 1 laying armour wires around cone

(e) Engage items 2 and 1 and tighten up.

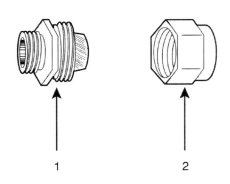

1   2

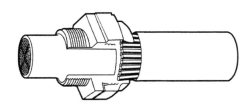

For use with all types of PVC-SWA cable where no seal is required onto cable inner or outer sheathing.

CW GLAND with OUTER SEAL

Indoor or outdoor gland for plastic or rubber insulated, single wire armoured, plastic or rubber oversheathed cables with extruded or lapped bedding.

Application

For use under most climatic conditions. CW glands are weatherproof and waterproof, and may be used in corrosive conditions when protected by a PCP shroud.

Specification

Gland consists of: gland body with knurled integral armour clamping cone (1), gland barrel assembly (2) including gland nut, captive PCP outer sealing ring and skid washer.

All metal parts are made of brass, accurately machined for ease of assembly.

The cable oversheath is sealed by tightening the gland nut; the washer allows the gland nut to 'skid' during assembly, completely eliminating the risk of torque being induced in the cable sheath.

Fitting sequence

As for previous gland except after engaging items 2 and 1 and tightening, the captive gland nut is tightened up.

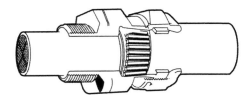

For use with all types of PVC-SWA cables where it is essential to produce a moisture-proof seal to IP66 onto the cables sheathing over armour.

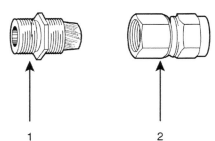

1   2

E1W GLAND with DOUBLE SEAL
(with inner and outer seals)

Indoor or outdoor gland for plastic or rubber insulated, single wire armoured, plastic or rubber oversheathed cables with extruded bedding.

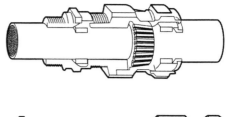

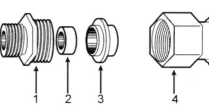

Application

For use in most climatic conditions and where it is essential to produce a seal to IP66 onto cable sheathing over cores and cable sheathing over armour, e.g. food industry etc., where jets of water are used for cleaning. E1W glands are weatherproof, and may be used in corrosive conditions when protected by a PCP shroud.

Specification

Gland consists of: gland body (1) polychloroprene (PCP) inner sealing ring (2), armour clamping cone (3), gland barrel assembly (4) including gland nut, captive PCP outer sealing ring and skid washer.

The inner seal is compressed to form a pressure tight seal with the extruded cable bedding, and the gland design prevents overtightening of the seal when clamping the armour wires. The cable oversheath is sealed by tightening the gland nut; the washer allows the gland nut to 'skid' during assembly, completely eliminating the risk of torque being induced in the cable sheath.

All metal parts are made of brass, accurately machined for ease of assembly.

Fitting sequence

(a) Screw item 1 into apparatus

(b) Pass item 4 over cable before commencing to strip oversheath

(c) Cut armour to length, lift wire ends, pass item 3 over exposed bedding and beneath armour

(d) Pass cable end through item 1

(e) Engage items 4 and 1 and tighten

(f) Unscrew item 4 and remove with cable end from item 1

(g) Check armour clamp

(h) Pass item 2 over exposed bedding

(i) Pass cable end through item 1

(j) Re-engage items 4 and 1 and tighten

(k) Tighten captive gland nut.

INSTALLATION PRACTICES TO BS 7671

COMMON RULES *(Chapter 51)*

General *(510)*

All items of equipment used in an electrical installation must be selected and erected so as to comply with the requirements of the regulations.

Compliance with Standards *(511)*

Every piece of equipment used in an electrical installation should comply with the requirements of the latest edition of the relevant British Standard or Harmonised European Standard.

Note: A list of publications of the British Standards Institution to which reference is made in the IEE Regulations is given in Appendix 1 of the regulations. For further details of BSI publications, reference should be made to the BSI Yearbook.

Operational Conditions and External Influences *(512)*

All equipment must be suitable for the:

- nominal voltage U_o (rms value for AC)
- design current
- current likely to flow in abnormal conditions (the duration of which is dependant on the characteristic of relevant protective devices)
- frequency
- power characteristics

of the installation or part of the installation concerned.

Any equipment installed should be selected and erected so that it is compatible with all other equipment in the installation; and be of suitable design for the environment in which it is to function and other conditions likely to be encountered including the test requirements of Part 7.

Socket outlets installed near sink tops

Standard BS 1363 socket outlets should not be installed too close to sink tops where they will be splashed with water or likely to be operated by persons with wet hands.

As a general guide in domestic premises, standard BS 1363 socket outlets or other similar accessories should be installed at least 300 mm (ideally further) horizontally from the edge of the sink top.

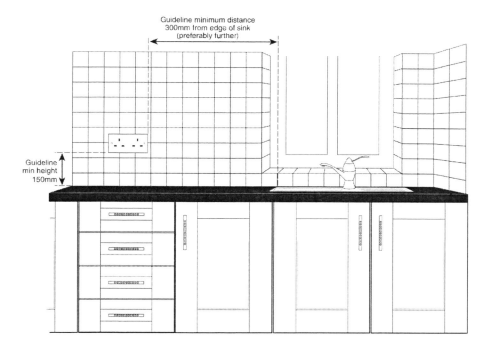

If it is known that an accessory will be subject to splashes of water, then equipment to at least IPX4 (splash proof) should be installed.

The height of a socket outlet above a worktop should be sufficient to minimise any mechanical damage that could occur to the socket or plug and flex during insertion, use or removal of the plug from the socket outlet (as a general guide, 150 mm minimum height).

Another example of a situation where external influences might affect the choice of equipment and conductors in a domestic dwelling is where the general temperature in the premises is unlikely to exceed 30°C, but may be considerably higher in airing cupboards or other localities. It is necessary to check that accessories, cables and other equipment affected are suitable for use at the higher temperature.

Accessibility *(513)*

All equipment should be installed so that it can be easily operated, inspected and maintained and provide ease of access to any connections. This regulation does not apply to joints in cables where such joints are permitted to be inaccessible (526-04).

Identification and Notices *(514)*

An installation must be labelled or identified by a suitable method to indicate its purpose.

All wiring shall be arranged or marked so it can be readily identified for inspection, testing, repairs or alterations.

Except where there is no possibility of confusion, clear marking must be provided at the connection of conductors identified in accordance with the harmonised cable core colours of BS 7671: 2004 and conductors identified to earlier versions of the regulations. A notice in accordance with 514-04 must be displayed (see BS 7671, Appendix 7).

Identification of Conductors *(514-03)*

Every cable core should be identified at its terminations and preferably throughout its length. Methods of identification may include coloured insulation applied to conductors during manufacture or the application of binding or sleeves (to BS 3858). The colours used must be those specified for the function in Table 51, BS 7671.

Note: Table 51 covers the identification of both fixed wiring and flexible cords and cables.

Cores of cables should be identified by:

- colour
- lettering and/or numbering

An example of identification of conductors and core colours to Table 51, BS 7671 is shown below.

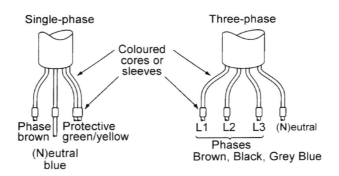

Switchboards

Identification of Conductors by Colour *(514-04)*

For identification by colour, a neutral conductor must be BLUE.

Protective Conductors

The two-colour combination of GREEN and YELLOW is exclusively used to identify a protective conductor. This colour combination must not be used for any other purpose.

GREEN and YELLOW single core cables should only be used as protective conductors and must not be over marked at their terminations.

Identification of a Protective Device *(514-08)*

All protective devices in an installation should be arranged and identified so that their respective circuits may be easily recognised.

Diagrams *(514-09)*

Diagrams and charts must be provided for every electrical installation indicating:

- The type of circuits
- The type of wiring system
- Details of the characteristics of the protective devices
- Circuits or equipment vulnerable to a typical test

Note: For simple installations, the above information may be given in a schedule. If symbols are used, they should conform to BS EN 60617.

A typical chart and diagram for a small installation are shown here.

Schedule of installation at ....

Type of circuit	Points served	Phase conductor mm²	Protective conductor mm²	Protective devices	Type of wiring
Lighting	10 downstairs	1 mm²	1 mm²	6 amp Type B circuit breaker	PVC/PVC
Lighting	8 upstairs	1 mm²	1 mm²	6 amp Type B circuit breaker	PVC/PVC
Cooker	1 Kitchen	6 mm²	4 mm²	32 amp Type B circuit breaker	PVC/PVC
Immersion Heater	Landing	2.5 mm²	1.5 mm²	16 amp Type B circuit breaker	PVC/PVC
Ring	10 downstairs	2.5 mm²	1.5 mm²	32 amp Type B circuit breaker	PVC/PVC
Ring	8 upstairs	2.5 mm²	1.5 mm²	32 amp Type B circuit breaker	PVC/PVC
Shower	Bathroom	10 mm²	4 mm²	40 amp Type B circuit breaker	PVC/PVC

Inspection and Testing *(514-12)*

Upon completion of an electrical installation, the electrical contractor should fix a label with details of the date of the last inspection and the recommended date of the next inspection. This label should be fixed in a prominent position, on or near the origin of the installation.

The notice must be inscribed with characters (not smaller than 11 point), as illustrated below.

IMPORTANT

This installation should be periodically inspected and tested, and a report on its condition obtained, as prescribed in BS 7671 requirements for Electrical Installations published by the Institution of Electrical Engineers.

Date of last inspection ....

Recommended date of next inspection ....

Residual Current Device – Notices *(514-12-02)*

When an installation incorporates a residual current device, a notice must be fixed in a prominent position, at or near the main distribution board. It should be printed in indelible characters, not less that 11 point in size and should read as follows:

> This installation, or part of it, is protected by a device which automatically switches off the supply if an earth fault develops. Test quarterly by pressing the button marked "T" or "Test". The device should switch off the supply and should then be switched on to restore the supply. If the device does not switch off the supply when the button is pressed, seek expert advice.

Earthing and Bonding *(514-13)*

A warning notice (as illustrated below) must be fitted in a visible position near to the point of connection of an earthing conductor to an earth electrode, a bonding conductor to an extraneous conductive part and the main earth terminal when it is separate from main switchgear.

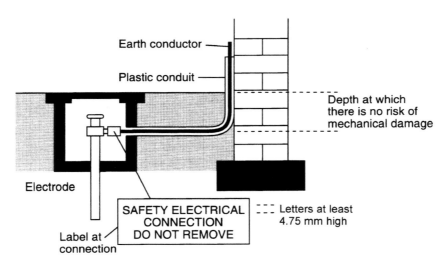

Warning Notice – Non-standard Colours *(514-14)*

If alterations or additions are made to the wiring of an installation so that some wiring complies with the current version of BS 7671 (the harmonised cable core colours), but the other wiring complies with an earlier version of BS 7671, a warning notice should be fixed at or near the distribution board concerned stating:

> **CAUTION**
>
> This installation has wiring colours to two versions of BS 7671. Great care should be taken before undertaking extension, alteration or repair that all conductors are correctly identified.

Mutual Detrimental Influence *(515)*

All electrical equipment should be selected and erected so as to avoid any harmful influences between the electrical installation and any non-electrical services. When equipment carrying currents of different types (AC or DC) or at different voltages is grouped in a common assembly, all equipment using any one type of current or any one voltage must be effectively segregated from equipment of any other type, to avoid mutual detrimental influence.

CABLES, CONDUCTORS AND WIRING MATERIALS *(Chapter 52)*

Selection of Type of Wiring System *(521)*

Cables and Conductors for Low Voltage

- Flexible or non-flexible cables or flexible cords used on low voltage shall comply with British or Harmonised Standards

Flexible Cords

Flexible cords are up to 4 mm² in size (flexible cable $\geq$ 4 mm²) and should be selected in accordance with Table 4H3A, Appendix 4, IEE Regulations.

Flexible cords must not be used for fixed wiring unless contained in an enclosure providing protection against mechanical damage (521-01-04) or relevant provisions of the regulations are complied with.

Flexible cords should be used for making the electrical connection to portable or fixed equipment; care should be taken to keep their exposed length as short as possible without undue strain on the conductor or sheath, by making the connection to fixed wiring using suitable accessories such as plugs, sockets, joints, boxes, etc. (526)

Where luminaires are suspended from flexible cords the maximum weight supported must not exceed the value in Table 4H3A, IEE Regulations. For example; for 0.5 mm² flexible cord the maximum which can be supported would be 2 kg; for a 1 mm² flexible cord it would be 5 kg.

Electromechanical Stresses *(521-03)*

All conductors and cables must be adequate for their purpose and installed so as to be able to withstand the electromechanical forces created by any current, including fault current that they may have to carry.

Lighting Track *(521-06)*

Any lighting track systems should comply with BS EN 60570.

Methods of Installing Cables and Conductors *(521-07)*

BS 7671 specifically provides for the installation methods of Table 4A1. Other methods are allowed providing that compliance with BS 7671 is maintained.

Any non-sheathed cables must be in conduit, ducting or trunking (this does not apply to protective conductors complying with Section 543).

External Influences *(522, see also Chapter 32 and Appendix 5)*

From Appendix 5 of BS 7671, external influences are described by the use of two capital letters and a number.

The first letter indicates the general category as follows:

    A = Environment

    B = Utilisation (who is using the building, materials are stored or processed, etc.)

    C = Building construction

The second letter refers to the type of influence, for example, (AD) Water.

The number describes the class, for example, (AD4) Splashes.

Ambient Temperature (AA) *(522-01)*

The type of conductor, cable, flexible cord and joint used in the wiring of circuits, must be suitable for the highest operating temperature likely to occur in normal service. Account should be taken of the minimum temperature likely to occur so as to avoid the risk of mechanical damage to those cables susceptible to low temperatures, such as PVC insulated cables, which crack if installed in refrigeration plants where the temperature is consistently below freezing point.

External Heat Sources *(522-02)*

In order to avoid the effects of heat from external sources (such as solar gain) from the sun, any of the following actions should be taken:

- shielding
- placing at a distance
- selecting a suitable wiring system
- reinforcing or substituting the insulating material

Where cables or flexible cords are in contact with equipment or accessories which transfer heat, such as immersion heaters and luminaires, termination to this equipment should be made using heat-resisting flexible cable or cord, or a suitable supplementary insulated sleeve or insulation should be applied to the conductors *(522-02-01)*.

Enclosures should be selected so that they are suitable for the extremes of ambient temperature that will be encountered in normal service. A typical example might be PVC conduit which distorts in hot weather if expansion couplings have not been correctly fitted.

Presence of Water (AD) or High Humidity (AB) *(522-03)*

Wiring systems should not ideally be exposed to rain, water or condensation, but where this cannot be avoided, the wiring system should be selected so that no damage is caused.

Dust, Solid Foreign Bodies (AE) *(522-04)*

Wiring systems should be installed to minimise the ingress of solid foreign bodies. Where dust or similar may accumulate, precautions should be taken to prevent any adverse effects on the heat dissipation of wiring or equipment.

Corrosive and Polluting Substances (AF) *(522-05)*

Where a corrosive or polluting environment cannot be avoided the wiring system used should be of a type (or be protected) so as to withstand exposure to the corrosive or polluting substances. Non-metallic materials should not be placed in contact with oil/creosote or similar hydrocarbon substances likely to cause chemical deterioration.

There should be no contact with materials likely to cause electrolytic action, deterioration or hazardous degradation.

Materials likely to cause corrosion of wiring systems are:

- materials containing magnesium chloride (used in the construction of floors and dados)
- plaster undercoats containing corrosive salt
- lime, cement and plaster
- oak and other acidic woods
- dissimilar metals liable to set up electrolytic action, e.g. copper and aluminium

Where conductors require termination involving soldering (as in the sweating on of cable lugs), the soldering flux used must not remain acidic or corrosive after the completion of the soldering process.

Polystyrene

Thermoplastic (PVC) cables must be separated from expanded polystyrene materials (e.g. polystyrene granule insulation) to prevent the plasticiser migrating to the polystyrene. This makes the cable sheathing less flexible and the polystyrene becomes soft and sticky.

Timber Treatment

PVC cables should not come into contact with wood preservatives during their application until any solvents have evaporated. Also, creosote should never be allowed near PVC cables as it will cause decomposition of the cable (e.g. swelling and loss of flexibility).

Impact (AG) *(522-06)*

Wiring systems should be designed to minimise mechanical damage, i.e. impact, abrasion, penetration, compression or tension.

The practice of covering cables with plaster is widespread and cases do occur of nails and other objects penetrating cables and causing damage. This gives rise to the risk of electric shock. In practice, although this risk seems to be small, it is desirable to reduce the risk as much as possible.

In an installation where a medium or high severity impact could occur then protection can be provided by one or more of the following:

- mechanical characteristics of the wiring system
- location
- additional mechanical protection fitted locally or generally

Cables Installed Under Floors

Cables installed under floors and over ceilings must be routed so that they will be undamaged through contact with the floor or ceiling, or by the method of fixing. This involves careful routing and clipping of cables.

Cables run in the space under floors and over ceilings should be installed at least 50 mm below the surface to prevent penetration by the nails or screws used in fixing flooring and ceiling materials. Alternatively, cables should be installed in an earthed steel conduit that is securely supported or provided with equivalent mechanical protection that will prevent penetration by nails or screws, etc. (522-06-05).

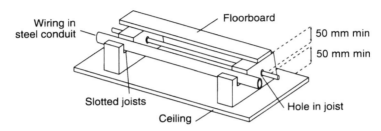

Support and protection for cables run under floorboards

Cables Concealed in Walls or Partitions

Regulation 522-06-06 requires that where cables are installed in walls or partitions at depths of less than 50 mm, the risk of damage is minimised. The permitted methods are as follows:

- Cables installed in a 150 mm zone from the top of a wall or partition, or within 150 mm of an angle created by adjoining walls or partitions.

- Cables may be run horizontally or vertically to accessories installed on walls or partitions. Note: If the location of an accessory can be determined from the reverse side and the wall or partition is of ≤100 mm thickness, the zone extends to the reverse side of the wall or partition.

In view of the practical problems, it is likely that cables will be installed mainly in the permitted zones.

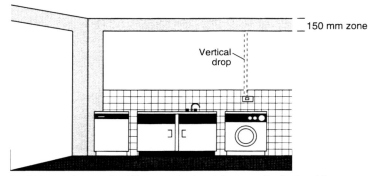

Cables should be run in permitted Zones or horizontally or vertically direct to accessory

Capping (metal or plastic) is used to protect the cable during the plastering operation, but gives very limited protection against nails and other objects driven into the plaster. Its use does **not** give compliance with 522-06-06.

Other Mechanical Stresses (AJ) *(522-08)*

Conductors and cables should be installed so that they are protected against any risk of mechanical damage.

Where cables pass through holes in metalwork, such as metal accessory boxes and luminaires, bushes or grummets must be fitted to prevent abrasion of the cables on any sharp edge.

Conductors and cables should not be subject to damage from incorrect bend radius, inadequate support, including damage from its own weight or be subject to any excessive mechanical strain. Flexible wiring should not suffer from undue torsional or tensile stresses on terminations or conductors.

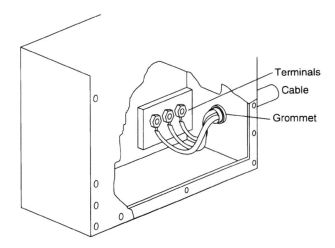

Flora and/or Mould Growth (AK) *(522-09)*

Where conditions due to vegetation, plants or mould growth present a hazard (AK2), the wiring system should be suitably selected or protected.

Fauna (AL) *(522-10)*

Wiring systems to be suitable or specially protected to avoid hazards (AL2) from insects, birds, small animals in harmful quantities or displaying an aggressive nature.

Solar Radiation and Ultraviolet Radiation (AN) *(522-11)*

Where cable and wiring systems are installed in locations that will be subject to significant amounts of solar radiation (AN2) or ultraviolet radiation, a suitable wiring system must be used or adequate shielding must be provided.

Equipment subjected to ionising radiation may need the application of special precautions.

Building Design (CB) *(522-12)*

If structural movement (CB3) occurs or is expected, cable supports and protection systems should be capable of permitting movement to prevent the conductors suffering excessive mechanical stress.

For flexible or unstable structures (CB4), a flexible wiring system must be installed.

Wiring systems must not penetrate load-bearing elements of buildings unless the integrity of the load-bearing element would be unaffected.

Index of Protection

When installing cables and accessories to meet the requirements of BS 7671 Chapter 52 section 522 gives classification and codes for each external influence, for example the installation of a socket outlet or luminaire. Outside where it is subject to being splashed by water. These classifications can be found in Appendix 5 of BS 7671.

Two of the most common codes are:

1. AE protection against foreign bodies such as a finger
2. AD protection against water

If were to install a luminaire on an outside wall the code would be AD4.

Product Marking

The requirements for marking products are specified in the relevant product standards, examples of 1 and 2 above are:

1. Protection against touching a live part in an electrical accessory with your finger would be IP4X.

2. Protection against splashes which could be caused by rain would be IPX3.

Current-Carrying Capacity of Conductors *(523)*

Cables selected should be capable of carrying the maximum sustained current for the circuits supplied (having considered the temperature and the environment in which the cable is installed), which should not cause the conductor operating temperature given in the tables of Appendix 4 to be exceeded (see also Table 52B).

Conductors in Parallel *(523-02)*

Except for a ring final circuit, cables connected in parallel to supply a load or piece of equipment must be of the same construction, cross-sectional area, length and disposition, without branch circuits and arranged as to carry substantially equal currents.

Thermal Insulation *(523-04)*

Examples of cables run in insulation are given below.

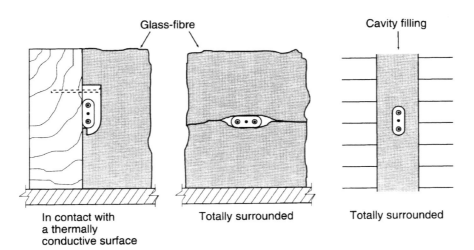

In contact with a thermally conductive surface | Totally surrounded | Totally surrounded

Where it is impossible to avoid running cables under the above conditions, a derating factor as illustrated below must be used. The current-carrying capacity of cables likely to be totally surrounded by insulation for lengths greater than 0.5 m must be reduced by 0.5 times the current rating for that cable when clipped direct (Ref. Method 1 in BS 7671 Appendix 4). For cables up to 10 mm², refer to IEE Regulations Table 52A.

When a cable, installed in a thermally insulated wall or above a thermally insulated ceiling is in contact with a thermally conductive surface allowing heat to be dissipated on one side, the cable current-carrying capacities given in BS 7671 Appendix 4, Method 4, 6 or 15 (as appropriate) are to be used.

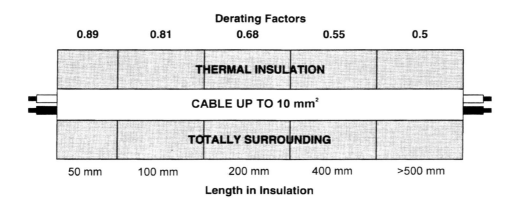

| E | **Resistance to Passage of Sound** |

New Building and those affected by a change of use, for example a loft conversion, are covered by Approved Document E.

The requirement is for adequate means of protection against sound for those people who live in dwelling houses, flats and rooms for residential purposes, which includes:

- protection against sound within a dwelling - house covering between internal floors and walls

Internal Floors

Many materials used for sound absorption often have good thermal insulation properties, e.g. mineral wool which could introduce risks for electrical installation work involving:

- reduced current-carrying capacities of cables covered by the material
- fire risk form a recessed luminaire mounted in a ceiling that is covered by the material

Care must be taken when selecting luminaries, especially recessed types, to ensure only those that have been tested and shown no to reduce sound insulation are used. Care must also be taken to ensure the fire integrity of the ceiling.

Neutral Conductors *(524-02)*

In a single-phase circuit, the neutral conductor must have the same csa in mm^2 as its associated phase conductor.

Voltage Drop *(525)*

The size of bare conductors or cables should be such that the voltage drop within the installation does not impair the safe function of connected equipment (in normal service). See relevant equipment British Standard.

The regulations will be satisfied if, for a supply complying with the Electricity Safety, Quality and Continuity Regulations 2002, the volt drop between the origin of the installation (normally the supply terminals) and a socket outlet or the terminals of fixed current-using equipment does not exceed 4% of the nominal voltage of supply.

Note: On installations supplying electric motors, account should be taken of the effect of the motor starting currents on other equipment

Electrical Connections (526)

Every connection of a conductor must be electrically and mechanically sound.

When deciding on the method of connection, the following should be considered:

- conductor material/insulation
- conductor size, number and shape of wires
- the number of conductors being connected together
- the conductor insulation should not be affected by the temperature of the terminations in normal use
- soldered connections should be suitable for creep, mechanical stresses and the temperature rise of fault current conditions
- protection against vibration

For all terminations made in enclosures, the enclosure should provide sufficient mechanical and external influence protection.

Every termination or joint in a live conductor should be within one (or more) of the following:

- an accessory or enclosure complying with British Standards
- an enclosure made or completed with non-combustible building material to BS 476-4
- an enclosure made or completed by the building structure should have ignitability characteristic 'P' as BS 476 (Part 5)

Note: A material having ignitability characteristic 'P' to BS 476 (Part 5) has been subjected to test where it has been placed in front of the flame of a gas jet for 10 seconds. If the material does not flame and burning does not extend to the edge within this time, it is classified as 'not easily ignitable' and its performance indicated by the letter 'P'.

Accessibility *(526-04)*

Every connection and joint must be accessible for inspection and testing and maintenance:

- joints made by welding, soldering, brazing or compression tool
- a joint forming part of the equipment (complying with product standards)

Proximity to Non-Electrical Services *(528-02)*

Electrical circuits and exposed metalwork

A wiring system installed close to a non-electrical service must be:

- suitably protected from any likely hazards possible from the other service in normal use, AND
- have indirect contact protection in accordance with Section 413

Heat, smoke or fumes *(528-02-02)*

A wiring system must not be installed in the vicinity of a service which produces heat, smoke or fumes detrimental to the type of wiring used, e.g. PVC cables close to balanced flue boilers.

Condensation *(528-02-03)*

A wiring system should be suitably protected when installed near a service which could cause condensation, e.g. water, steam, gas, etc.

Wiring systems installed near non-electrical services should be situated so that any work carried out on either service would not harm the other.

Maintainability *(529)*

Any protective measures that must be removed to allow maintenance to be carried out shall be reinstated where practicable without reducing the original degree of protection.

Safe and adequate access shall be provided to all parts of the wiring system that require maintenance (529-01-02).

B THERMAL EFFECTS

PROTECTION AGAINST THERMAL EFFECTS *(Chapter 42)*

The regulations in this chapter are intended to deal specifically with the protection of persons, fixed equipment and fixed materials next to electrical equipment from:

- combustion, ignition or damage to materials
- burns
- impaired safety of installed equipment

Electrical equipment must not present a fire hazard to any adjacent materials.

Most sections of the regulations contribute in some way to protection against thermal effects, e.g. proper selection of type and size of cables, segregation, fire barriers, methods of installation, correct selection of overcurrent devices.

Protection Against Fire *(422)*

In the case of adjacent materials which may be a fire hazard, or equipment installed where operating surface temperature may give rise to the risk of fire, one or more of the following methods must be adopted:

- mounting on or enclosed in, or screened by materials which will withstand the temperature generated, without the risk of fire or harmful effects
- screened by heat-resistant materials
- mounted to allow heat to dissipate safely

Where arcing or high temperature particles could be emitted from fixed equipment, one or more of the following methods should be used:

- enclosed in arc-resistant material
- screened by arc-resistant material
- mounted at a suitable distance from any material which could be harmfully affected, allowing the safe extinction of any emissions

Fixed luminaires and lamps should be guarded to prevent ignition of any materials which are likely to be placed in close proximity. Shades or guards must be suitable to withstand the heat generated from the lamp or luminaire.

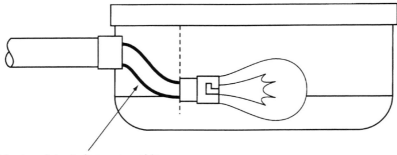

Heat resistant sleeve or cable

B | Recessed Luminaires

When recessed luminaires are installed in a ceiling which is a fire barrier they should be selected and installed so they do not reduce the fire integrity of the ceiling.

This can be achieved in some cases by using a fire hood which is placed above the ceiling over the luminaire. The term used to describe this hood is an intumescent cover. These cover when used should not restrict the ventilation to the luminaire causing it to overheat and they need to be compatible with luminaire.

B | Lighting Diffuses

Thermoplastic lighting diffuses forming part of a ceiling, for example recessed luminaries, the selection and installation of these diffuses must conform to the Building Regulations.

The requirements are for situations where the luminaire is recessed into the ceiling and the thermoplastic diffuses is part of the ceiling. This does not apply to diffuses of luminaries attached to the soffit of a ceiling or suspended beneath it.

Thermoplastic lighting diffuses are classified as:

- a translucent or open-structured element that allows light to pass through. It may be a part of a luminaire or used below a roof light or other source of light

These lighting diffuses may form parts of ceilings to rooms without restrictions on use providing the following conditions are met:

- the diffuses are of classification TP(a) rigid

- if the diffuses classification is TP(a) flexible or TP(b) their use in limited as specified in Table II of Approved (Building Regulations) Document B

L Approved Document L1 of the Building Regulations (England and Wales) requires that reasonable provision be made to conserve fuel and power in the following cases:

- new dwellings
- extensions to existing dwellings
- loft conversions
- new outdoor lighting installations

For internal lighting points pendants or a luminaire should be installed which can only take energy efficiency lamps having a luminous efficiency greater than 40 lumens per circuit watt which should include any associated control gear.

To ensure the lighting fittings or luminaries continue to only operate using energy efficient lamps, they must not be capable of using GLS tungsten lamps.

In the case of external lighting fixed to the dwelling, including porches, energy efficient lamps having a luminous efficiency of greater than 40 lumens per circuit watt should be used in the selected luminaire, together with a means to automatically switch of the luminaire when there is enough daylight (photoelectric cell) and when not required (movement sensor).

Protection Against Burns *(423)*

Any part of an enclosure of fixed equipment liable to reach a temperature that would cause burns (i.e. in excess of the values in Table 42A) must be located or guarded so as to prevent accidental contact. An exception is made for any equipment manufactured to a British Standard which has a specific limiting temperature.

Equipment within arm's reach must comply with Table 42A of the IEE Regulations.

Typical examples:

Metallic hand-held equipment	55°C
Non-metallic hand-held equipment	65°C
Metallic equipment which can be touched	70°C
Non-metallic equipment which can be touched	80°C
Metallic equipment that does not need to be touched	80°C
Non-metallic equipment that does not need to be touched	90°C

Protection Against Overheating *(424)*

The electric heating elements of forced air heating systems (other than central storage heaters) must not be activated until the required air flow has been achieved, and should be de-activated if the air flow stops or reduces.

Also, two independent temperature limiting devices should be installed to prevent permissible temperatures being exceeded.

Frames and enclosures of electric heating elements must be of non-ignitable material.

Electric appliances that produce hot water or steam must by their design or erection method be protected against overheating in all service conditions.

CHOICE OF PROTECTIVE MEASURES AS A FUNCTION OF EXTERNAL INFLUENCE *(Chapter 48)*

These requirements are in addition to Chapter 42 and Section 527 and apply to:

- locations with risks of fire due to the nature of processed or stored materials
- installations in areas built from combustible materials

Electrical equipment must be selected and erected so that its normal operating temperature (or any temperature rise due to fault) will not cause a fire. This can be achieved by the equipment construction or the provision of additional protection.

Locations With Risks of Fire Due to The Nature of Processed or Stored Materials *(482-02)*

All equipment must have a degree of protection of at least 1P5X (482-02-03).

Cables not completely embedded in non-combustible materials (e.g. plaster) or protected from fire by other means must meet the requirements of BS EN 50265-2.1 or 2.2 for flame propagation.

In the absence of other recommendations from manufacturers, spotlights or projectors should be installed at a distance from combustible materials of at least:

- lamp rating up to 100 W   –   0.5 m
- lamp rating 100 W–300 W   –   0.8 m
- lamp rating 300 W–500 W   –   1.0 m

Any luminaires should be of a design that prevents lamp components falling from them.

Any heating appliances must be fixed and, if installed close to combustible materials, barriers must be provided to prevent against ignition. Storage heaters must be of a suitable type to prevent any ignition of dust or fibres by the heat store.

Enclosure temperatures of heaters, etc. should not exceed 90°C under normal conditions and 115°C under fault conditions.

Locations with Combustible Construction Materials *(482-03)*

Electrical equipment which is installed in, or on a combustible wall, e.g. distribution boards, installation boxes etc., must comply with relevant standards for enclosure temperature rise (482-03-01). Equipment that does not comply with the above must be enclosed in suitable non-flammable material (with the effects of heat dissipation taken into account).

Cables and cords must comply with BS EN 50265 2.1 or 2.2.

Conduit and trunking must meet the fire resistance tests of, and comply with, BS EN 50086-1 and BS EN 50085-1.

Minimising the Spread of Fire *(527)*

Sealing of the Wiring System *(527-02)*

By the selection of appropriate materials that are suitably installed, the risk of spread of fire can be minimised.

The wiring system should not affect building structure performance or fire safety by the installation methods utilised.

Cables to BS 4066-1 may be installed without special precautions except where the risk of fire is high.

Cables not complying with BS 4066-1 should be only short lengths for connection of appliances to the fixed wiring system but they should not pass from one fire segregated compartment to another.

Conduit and trunking complying with the flame propagation requirements of BS EN 50085 or 50086 may be installed without special precautions.

A wiring system should be enclosed in non-combustible building materials having ignitibility characteristic 'P' as BS 476-5, if the materials of the system conform to a British Standard which has no requirement for testing for resistance to the propagation of flame.

Where a wiring system is required to pass through or penetrate material forming part of the construction of a building (e.g. cable trunking or busbar trunking systems), areas external to the wiring system, and where necessary, internal areas, holes must be sealed to maintain the required fire resistance of the material.

The sealing system used must meet the following requirements:

- be compatible with the wiring system concerned
- permit thermal movement of the wiring system without detriment to the sealing quality
- be removable without damage when additions to the wiring system are necessary
- be capable of resisting external influences to the same standards as the wiring system

During the installation of wiring systems, temporary sealing arrangements must be made. In addition, any existing sealing which is disturbed or removed in the course of alterations to an installation, must be reinstated as soon as possible.

[B] It is essential that sealing arrangements are visually inspected during installation to verify that they conform to the manufacturer's instructions. Details of those parts of a building sealed and the methods used must be recorded.

Any joints between the fire separating parts of a building should be fire stopped to prevent the spread of fire.

Any openings for pipes, ducts, conduits or cables which pass through fire separating parts of a building such as walls, floors and ceiling should be limited in number and kept as small as practical and fire stopped.

Where cables pass through walls, floors and ceilings the fire stopping material should be incombustible material which maintains the fire resistance of the wall, floor or ceiling, examples are and should be used as appropriate:

- a proprietary fire stopping and grilling system
- cement mortar
- gypsum based plaster
- cement
- glass fibre
- intunslucent matics

[F] ## Ventilation

Approved Document F (England and Wales) applies to all new buildings and those created by a change of use.

The requirement is that there shall be adequate means of ventilation provided for persons in buildings.

Table 1 of section 1 provides details of extract ventilation rates for different locations where mechanical extract fans are installed.

The following publications contain additional advice on achieving compliance with ventilation requirements:

- BS 5720: 1979 Code of Practice for mechanical ventilation and air-conditioning of building

- BRE Digest 398 Continual mechanical ventilation in dwelling: design installation and operations

Note: When mechanical extract fans are to be installed where open flues appliances are located the appliance needs to operate safely:

- in the case of a gas appliance in a kitchen the maximum extract rate is 20 litres/second and BS 5440-1 should be referred to for details of carrying out a spillage test on the installed appliance

- for an oil-fired appliance the installation is to be carried out in accordance with the Oil Firing Technical Association's technical information note T1/112

- mechanical extract fans should not be installed in the same room as a solid fuel appliance

SPECIAL INSTALLATIONS AND OTHER EQUIPMENT

Introduction

Special installations or locations are now covered in separate sections of the regulations, the particular requirements for each of them supplement or modify the general requirements of BS 7671.

The absence of reference to the exclusion of a chapter, section or clause means the relevant general regulations apply, see Regulation 600-02.

The special installations or locations are:

Rooms containing fixed baths and showers   Section 601

Rooms Containing a Bath or Shower *(601)*

This section covers baths, showers (any associated cubicles) and their surroundings. It does not apply to emergency facilities in laboratories or industrial areas.

Areas that contain baths or showers for medical treatment or disabled persons may require special consideration.

Zone Classification

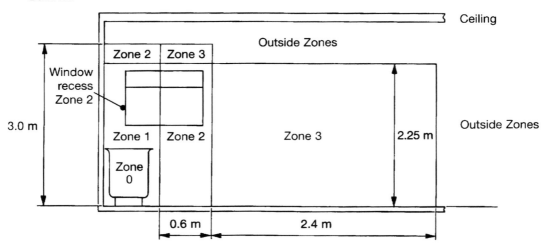

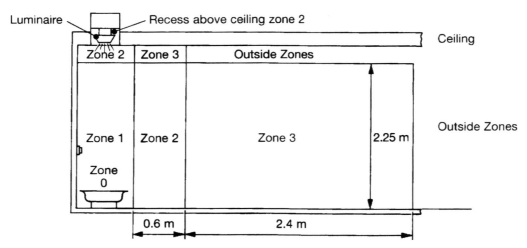

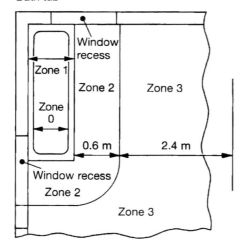

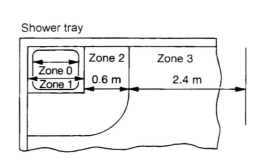

In the zone classification diagrams Zone 0 is the interior of the bath tub or shower tray.

The area below the bath or shower tray is classed as Zone 1 if access is without the use of a tool. If access is only by the use of a tool then this space is considered to be outside the zones.

Electric Shock Protection

Where SELV or PELV is used, direct contact protection shall be provided by:

- either, barriers or enclosures to IP2X or IPXXB
- or, insulation able to withstand a type test voltage of 500 V AC for one minute

Protective Measures Against Electric Shock

In Zone 0, only SELV not exceeding 12 V AC or 30 V ripple free DC is allowed with the safety supply source being installed outside Zones 0, 1 and 2.

The following methods are **not** permitted:

- protection by obstacles
- placing out of reach
- non-conducting location
- earth-free local equipotential bonding

External Influences

Electrical equipment shall have at least the following IP ratings:

Zone 0   IPX7

Zone 1&2   IPX4
IPX5 (where water jets are used for cleaning purposes)
This does not apply to shaver supply units to BS EN 60742 installed in Zone 2 where direct spray from showers is unlikely.

Zone 3   IPX5 (where water jets are used for cleaning purposes)

Wiring Systems

For surface wiring or wiring embedded in a wall up to 50 mm deep and which does not comply with 522-06-07:

Zone 0   Only wiring for fixed equipment installed in that zone

Zone 1   Only wiring for fixed equipment installed in Zones 0 or 1

Zone 2   Only wiring for fixed equipment installed in Zones 0, 1 or 2

Switchgear

These requirements do not apply to the controls of fixed equipment (suitable for use in a given zone) incorporated in the equipment:

Zone 0　　No switchgear or accessories

Zone 1　　Only switches for SELV (max 12 V AC or 30 V ripple free DC) with the safety supply source being installed outside Zones 0, 1 or 2.

Zone 2　　No switchgear, accessories with switches or socket outlets to be installed except for:

    1　Switches or socket outlets for SELV circuits (safety supply source must be installed outside Zones 0, 1 or 2)

    2　Shaver supply units to BS EN 60742

Zone 3　　No socket outlets except for:

    1　SELV socket outlets complying with Reg 411-02
       Shaver supply units to BS EN 60742

    2　There should be no means for connecting portable equipment except for items in 1 above

Only the insulated cord of a pullcord switch to BS 3676 is permitted in Zones 1 and 2. The body of the switch must be installed outside Zones 0, 1 and 2.

Except as allowed by 601-08-02, there must be no socket outlets other than SELV or shaver supply units to BS 60742 installed outside Zones 0, 1, 2 or 3.

Where a shower cubicle is installed in other than a bath or shower room then any socket outlets outside Zones 0, 1, 2 or 3 (other than a SELV socket or a shaver supply socket to BS EN 60742) must be protected by a 30 mA RCD.

Fixed Current-using Equipment

Zone 0:　Only fixed current-using equipment specifically designed for that zone to be installed

Zone 1:　If it is specifically designed for Zone 1, the following may be installed:

- water heaters

- shower pumps

- other fixed current-using equipment which can only be located in Zone 1 provided it is suitable for the conditions and further protected by a 30 mA RCD

- SELV current-using equipment

The following information does not apply to fixed equipment fed from a SELV supply to 411-02 or 601-03-02.

Zone 2: If it is specifically designed for Zone 2, the following may be installed:

- shower pumps
- water heaters
- luminaires, fans, heating appliances
- units for whirlpool baths (complying with relevant standards)
- other fixed current-using equipment that can only be located in Zone 2 (provided it is suitable for the conditions of the zone)
- SELV current-using equipment

Any non-fixed current-using equipment in Zone 3 must be protected by a 30 mA RCD.

Underfloor electric heating in any zone should be covered by an earthed metallic grid or have an earthed metallic sheath, the grid or sheath being connected to the local supplementary bonding (601-09-04).

SAFE ISOLATION

Introduction

Electricity, when safely controlled, is a very efficient and convenient way of distributing and using energy. If it is inadequately controlled, it can be lethal.

Electric Shock

This is an effect on the nervous system, with a resulting contraction of muscles and feeling of concussion.

An electric shock can be received either by direct or indirect contact.

Direct contact – contact of persons with parts or conductors that are intended to be live in normal use.

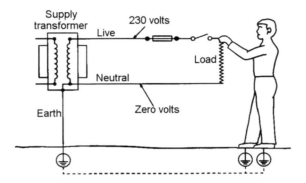

Indirect contact – contact of persons with exposed or extraneous conductive parts (metalwork) that have become live under fault conditions.

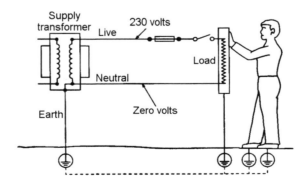

Body Resistance

This varies due to the conditions of the body, age, situation and weather.

Perception level

1 mA is the point at which an individual is aware of electric current.

Let Go Level

9 mA is the point at which an individual still has control over the effects of shock on the muscles in the body to be able to release a gripped conductor.

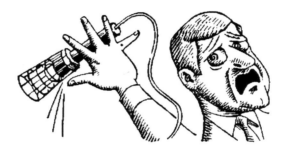

Freezing and Muscular Contraction

With further increases in levels of current (around 20 mA) the subject cannot release themselves. Extreme pain is felt which may cause the subject to lose consciousness, or the body muscles may contract affecting the lungs and the subject may die from asphyxia.

Death

With currents of around 80 mA, death is likely to occur by ventricular fibrillation and severe internal and external burns.

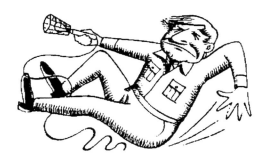

Ventricular Fibrillation

This is the effect of electric shock which causes the muscles of the heart to contract separately at different times instead of in unison. This condition is a killer.

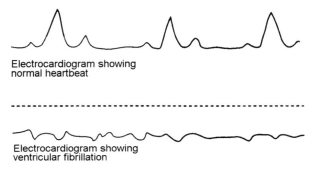

Removing Persons

Removing a person from contact with live conductors needs great care. The following points should be considered:

- the rescuer must not put themselves in danger
- procedures must be undertaken as quickly as possible
- procedures to be carried out in a way that prevents further injury

Disconnect the electricity supply first wherever possible. Pull victim away from live conductors using dry clothing such as overalls wrapped around them to enable them to be effectively and quickly removed.

Treatment

Summon assistance and call for an ambulance. In the case of slight shock, reassure and make the patient comfortable and report the accident to appropriate personnel.

If burns have been sustained, cool the areas with cold water or any other non-flammable fluid at hand. Remove anything of a constrictive nature if possible, such as rings, belts and boots. If the burns are serious, cool the areas and send the patient to hospital without delay.

For severe cases of shock where the patient is unconscious and not breathing, clear the airway and administer mouth-to-mouth resuscitation – remember there is no time to waste, as lack of oxygen to the brain can cause damage within four minutes.

On restoration of breathing, place the patient in the recovery position and send him/her to hospital without delay.

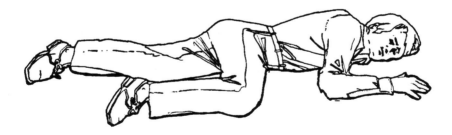

If the heart has stopped then cardiac compression should be given.

Electrical Isolation

Before beginning work on any electrical circuit you should make sure that it is completely isolated from the supply. Electrically powered machines are usually fitted with an isolator for disconnecting the supply under 'no-load' conditions. Otherwise, it may be necessary to isolate the supply by removing fuses, locking off MCBs or by physical disconnection of live conductors.

Fuses, switch-fuses and isolators should be clearly marked to indicate the circuit they protect.

Any isolating device, when operated, should be capable of being locked in the open position or carry a label stating otherwise. If the isolator consists of fuses, these should be removed to a safe place where they cannot be replaced without the knowledge of the responsible person concerned. For example, they can be kept in the pocket if the job is of short duration, or in a locked cupboard provided for the purpose in the charge of the works or site supervisor.

Fuses should never be removed or replaced without first switching off the supply.

Your Responsibility

When working on a particular circuit or appliance you must be certain that the supply cannot be switched on without your knowledge. You must satisfy yourself that the circuit is open and labelled so.

Accident history has shown that multimeters have frequently caused accidents when used for proving for the presence or absence of voltage. The preferred equipment to use is an approved voltage indicator or test lamp used in accordance with and complying with HSE guidance note GS38.

Voltage Indicators

Only approved voltage indicators and test lamps complying with and used in accordance with HSE GS38 should be used for establishing the presence or absence of voltage for safe isolation purposes.

These devices have a robust body containing the circuitry, internal resistors are fitted to limit the event of any fault occurring.

The exposed metal or the probe tip is kept to a maximum of 4 mm (ideally not more than 2 mm).

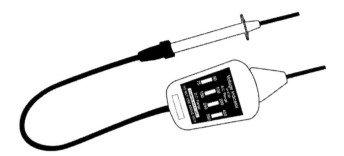

An approved voltage indicator

REMEMBER

These devices must be proved immediately before use and immediately after use on a known supply or purpose made proving unit (of same voltage as that under test).

Neon screwdrivers should not be relied on since the neon will not indicate supplies at low potentials. There is also the risk of receiving an electric shock if the resistor breaks down.

Flex and lampholders (home-made) test lamps should never be used. These are extremely dangerous, since mains potential is present in all the components; flex and bulb are vulnerable to damage and there is usually no fuse in circuit.

Local Isolation

It has been the practice of British Gas to ask for the electrical supply point, which will be used to supply a central heating system, to be installed as close as practical to the boiler and in a readily accessible position. The type of supply point being a 13 A BS 1363 un-switched socket outlet and plug top. The combination provides local isolation.

A more acceptable method of providing local isolation would be to install a double-pole switch fuse spur unit, in which the fuse carrier can be withdrawn but not removed and a small padlock fitted as illustrated.

When isolating a central heating unit, it is important to remember that the system will be electrically-connected to remote components within the property; therefore to ensure total isolation, all terminals at the appliance or wiring centre must be tested for the absence of electricity with all controls calling for heat. This will ensure, for example, that any temperature-controlled zoning system that has been wired from a separate power source will be identified and can be made safe before work is carried out on the system.

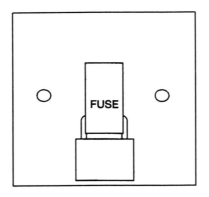

Isolation and Switching

The IEE Regulations state that a means of isolation is required for circuits and equipment in order to enable a skilled person to carry out work on that installation or equipment safely with it dead.

All isolating devices must comply with British Standards, which means the position of the contacts or other means of isolation must be externally visible or clearly indicated.

For TN-S and TN-C-S systems all phase conductors must be switched, and in TT systems all live conductors.

In a domestic or similar installation the Main Switch in a consumer unit must be double pole.

Functional Switching

The IEE Regulations also require a means of interrupting the supply on load for every circuit, for example a switch for a lighting point or immersion heater. Remember that in bathrooms insulated cord-operated switches should be used, the insulated pull cord being permitted in zones 1 and 2 but the switch body must be in zone 3.

SAFE ISOLATION PROCEDURE

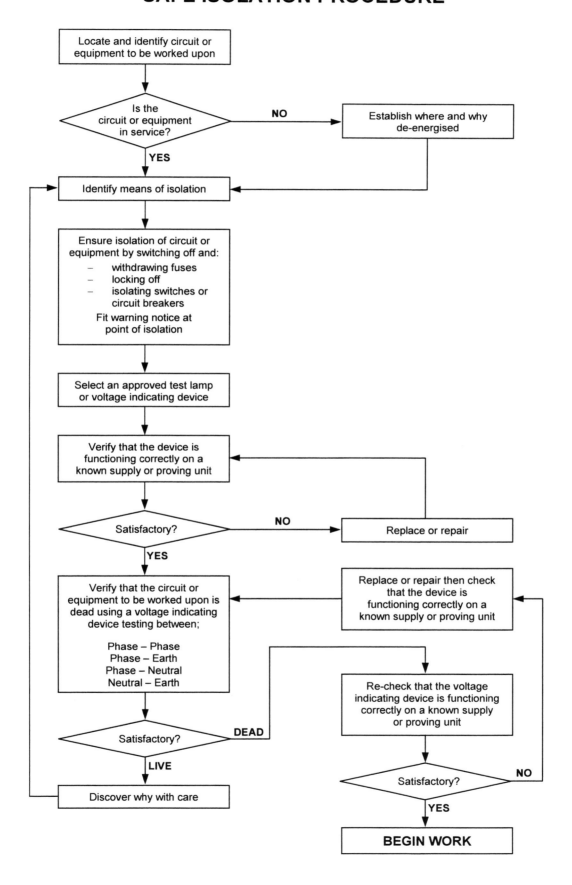

INSPECTION AND TESTING

General

Inspection should consist of a careful scrutiny of the installation, supplemented by testing to:

- verify safety of persons and livestock
- verify protection against damage to property by fire and heat
- establish that the installation is not damaged and has not deteriorated
- identify installation defects or departures from the regulations

Visual Checks

During installation and on completion, every installation must be inspected and tested before being connected to the supply to establish that the requirements of the regulations have been met. Precautions should be taken to avoid dangers to persons, damage to property and equipment during inspection and testing.

The following information should be made available to the persons carrying out the inspection and testing of an installation.

- Assessment of general characteristics:
  - maximum demand (after diversity)
  - the number and type of live conductors of the supply (and of the circuits of the installation), the earthing arrangements of the installation and any provision by the distributor
  - nominal voltage(s), nature of the load current and frequency
- Diagrams, charts or tables indicating:
  - the type of circuits
  - the number of points installed
  - the number and size of conductor
  - the type of wiring system
- Details of the characteristics of the protection devices for automatic disconnection, and a description of the method used for protection against indirect contact
- The location and types of devices used for:
  - protection
  - isolation and switching

- Details of circuits or equipment sensitive or vulnerable to tests – e.g. equipment with voltage-sensitive electronics, such as central heating controls with electronic timers and displays, also intruder alarm equipment, certain types of RCD, etc.

Note: For a new circuit the following information is also required:

- prospective fault current (PFC) at the origin of the circuit
- external loop impedance value of Ze
- type and rating of the protective device at the origin of the installation

Note: Information may be given in a schedule for simple installations. See example below for a domestic installation. A durable copy of the schedule relating to distribution board must be provided inside or adjacent to the distribution board.

Schedule of installation at..

Type of circuit	Points served	Phase conductor mm^2	Protective conductor mm^2	Protective devices	Type of wiring
Lighting	10 downstairs	1 mm^2	1 mm^2	6 amp Type B circuit breaker	PVC/PVC
Lighting	8 upstairs	1 mm^2	1 mm^2	6 amp Type B circuit breaker	PVC/PVC
Immersion Heater	Landing	2.5 mm^2	1.5 mm^2	16 amp Type B circuit breaker	PVC/PVC
Ring	10 downstairs	2.5 mm^2	1.5 mm^2	32 amp Type B circuit breaker	PVC/PVC
Ring	8 upstairs	2.5 mm^2	1.5 mm^2	32 amp Type B circuit breaker	PVC/PVC
Shower	Bathroom	10 mm^2	4 mm^2	40 amp Type B circuit breaker	PVC/PVC

Inspections

A detailed inspection should be made of installed electrical equipment, usually with the part of the installation being inspected disconnected from the supply. A Schedule of Inspection should be completed and signed and appended to the Electrical Installation Certificate or Periodic Inspection Report as appropriate.

The inspection should verify that it:

- complies with the British Standards or harmonised European Standards (this may be ascertained by mark or by certificate from the installer or manufacturer)
- is correctly selected and erected in accordance with these regulations
- is not visibly damaged so as to impair safety

The detailed inspection must include the following where relevant:

- connections of conductors
- identification of conductors
- routing of cables in safe zones or mechanical protection methods
- selection of conductors for current-carrying capacity
- connection of single-pole devices for protection or switching in phase conductors only
- correct connection of accessories and equipment
- presence of fire barriers, suitable seals, and protection against thermal effects
- methods of protection against electric shock
- protection against both direct and indirect contact, i.e.
  - SELV
- protection against direct contact:
  - protection by insulation of live parts
  - protection by barriers or enclosures
- protection against indirect contact:
  - earthed equipotential bonding and automatic disconnection of supply (EEBAD)
  - presence of earthing conductors
  - presence of protective conductors
  - presence of main equipotential conductors
  - presence of supplementary equipotential bonding conductors
  - use of Class II equipment or equivalent insulation
  - electrical separation
- prevention of mutual detrimental influence
- proximity of non-electrical services and influences
- separation of Band I and Band II circuits or Band II insulation used
- presence of appropriate devices for isolation and switching
- choice of protective and monitoring devices
- residual current devices
- overcurrent devices

- labelling of circuits, fuses, switches and terminals
- selection of equipment and protective measures appropriate to external influences
- adequacy of access to switchgear and equipment
- presence of danger notices and other warning notices
- presence of diagrams, instructions and similar information
- erection methods
- requirements of special locations

Note: During any re-inspection of an installation, all pertinent items in the checklist should be covered.

Testing

For initial verification, the following items (where relevant to the installation being tested) must be tested in the following sequence.

Before the supply is connected (or with the supply safely isolated):

- continuity of protective conductors including main and supplementary bonding
- insulation resistance
- polarity

With the electrical supply reconnected, carry out the live tests:

- earth loop impedance
- prospective fault current
- functional testing

The tests applicable should be carried out and the results compared to relevant criteria. The required tests up to, and including, polarity should be carried out in the stated order prior to the installation being energised.

If a test indicates failure to comply, that test and the preceding tests (whose results may have been affected by the fault) must be repeated after rectification of the fault.

Continuity of Protective Conductors

The initial tests applied to protective conductors are intended to verify that the conductors are both correctly connected and electrically sound. They also verify the resistance is such that the overall earth fault loop impedance of the circuits is of a suitable value to allow the circuit to be disconnected from the supply in the event of an earth fault (within the disconnection times selected to meet the requirements of Regulation 413-02-09).

Every protective conductor, including main bonding conductors and supplementary bonding conductors, should be tested to verify that the conductors are electrically sound and correctly connected.

Test methods 1 or 2, described on the following pages are used to confirm the continuity of protective and bonding conductors.

Test method 1, which measures the R1 + R2 value for the circuit being tested, may be the more convenient method to use as it will also confirm polarity as each test is completed. Test method 2 measures R2 only and has the inconvenience of the trailing wander lead.

To test the continuity of main and supplementary bonding conductors, only test method 2 can be used.

Test methods 1 and 2 can only be simply carried out on 'all insulated' installations.

Test instrument

Low-resistance ohmmeter with a recommended supply having a no load voltage of between 4 V and 24 V with a short circuit current of not less than 200 mA.

Note: Remember to subtract the value of the instruments test leads from the test results. Some instruments have an in-built facility to zero out or 'null' the test leads.

Test Method 1 ($R_1 + R_2$ Value)

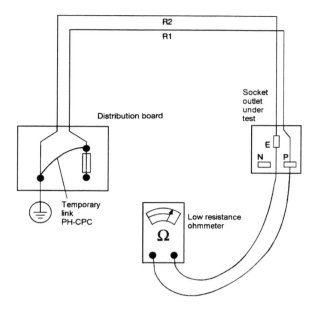

To carry out the test:

- safely isolate the supply

- temporarily link the phase and protective conductor at the supply end of the circuit, e.g. consumer unit or distribution board

- prepare the low-resistance ohmmeter for use, not forgetting to short the leads together and zero (null) the display or note the resistance of the test leads

- measure between the phase and protective conductor at each outlet, point or accessory on the circuit – this will also confirm polarity. Record the value obtained at the end of the circuit (max R1 + R2) on the Schedule of Test Results

    Note: Remember to deduct any test lead resistance if null facility not available

- remove the temporary phase-cpc link

Using this method will tell you a number of things about the circuit under test. It will:

- check the continuity of the phase and protective conductor

- verify that the polarity is correct at each test position

- provide a measure of the R1 + R2 value which, when added to the known value of impedance for the circuit distribution board (which would be Ze if the distribution board is at the origin), will give a calculated figure of Zs for the circuit which could be compared to a design value

Test Method 2 (R₂ Value)

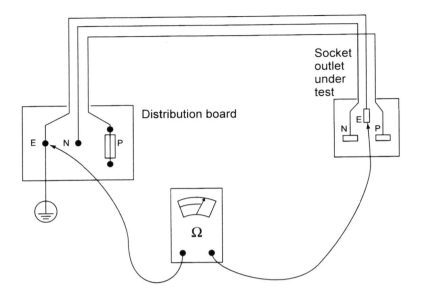

If the two ends of the protective conductor are within the span of the meter test leads, then the resistance of the protective conductor, R2, may be measured directly. If the distance between the two ends of the protective conductor exceeds the span of the instrument test leads, then they must be extended with a suitable length of cable.

Test method 2 is carried out by connecting one lead of the instrument to the main earthing terminal and using the other lead to make contact with all the protective conductors under test at the various points, accessories, outlets, exposed and extraneous conductive parts.

Note: The resistance of the extended test lead and instrument test leads must be deducted from the test results by either the 'null' feature on the instrument (if fitted) or measuring the overall test lead resistance and subtracting it from the results.

To carry out the test:

- safely isolate the supply

- a main or supplementary bonding conductor can be simply tested by disconnecting one end before testing

- prepare the low-resistance ohmmeter for use

- measure the resistance of the protective or bonding conductor(s) by connecting one lead from the instrument to the main earthing terminal (MET) and connecting the other lead to the various points under test

- remember to deduct or null the test lead resistance from your test results, and record the value obtained (R2) on the Schedule of Test Results

This test must be made to verify the continuity of the phase, neutral and protective conductors (unless the cpc is formed by conduit or trunking, etc.) of every ring final circuit.

Continuity of Ring Final Circuit Conductors *(713-03)*

This test must be made to verify the continuity of the phase, neutral and protective conductors (unless the cpc is formed by conduit or trunking, etc.) of every ring final circuit.

The test result should also establish that the ring is complete and has not been interconnected, creating an apparent continuous ring circuit that is actually broken. See illustration.

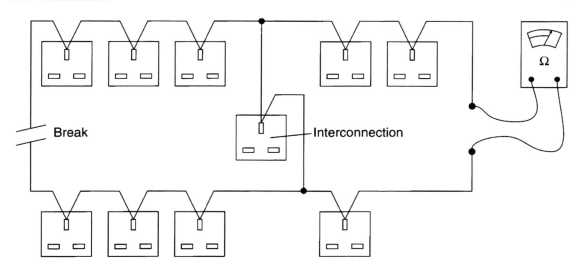

Test Instrument

Low-resistance ohmmeter

Step 1: Ring continuity

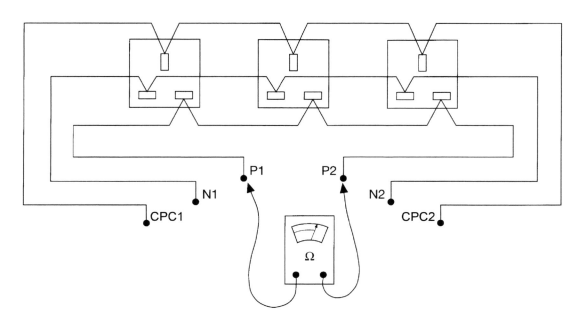

To carry out the test:

- safely isolate the supply

- identify and disconnect the phase, neutral and the circuit protective conductors at the distribution board or consumer unit, etc

- prepare the low-resistance ohmmeter for use

- measure the end-to-end resistance of each conductor. If the conductor size is the same, so should the instrument readings (within 0.05Ω). Any variation beyond this tolerance could be due to a wiring defect, loose connection etc. or incorrect identification of the ring conductors

Note: The values of the cpc ring will differ if the cpc is a smaller csa conductor than the phase or neutral, for example, 2.5/1.5 mm². In this case, the value of the cpc ring would be 1.67 times that of the phase or neutral ring, i.e. if the phase and neutral are both 1Ω, the cpc should be approximately 1.67Ω.

Step 2: Socket outlets, phase-neutral

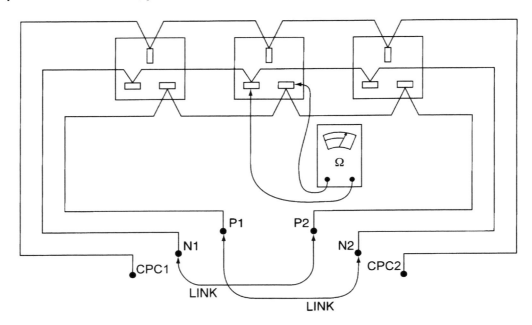

- Join the opposite ends of phase and neutral at the distribution board and measure between phase and neutral at every outlet or point on the circuit

- The values obtained should be substantially the same – approximately ¼ of the phase ring plus ¼ of the neutral ring resistance. Any sockets wired as spurs will have a higher reading in proportion to the length of the spur

Step 3: Socket outlets, phase-cpc (R1 + R2 value)

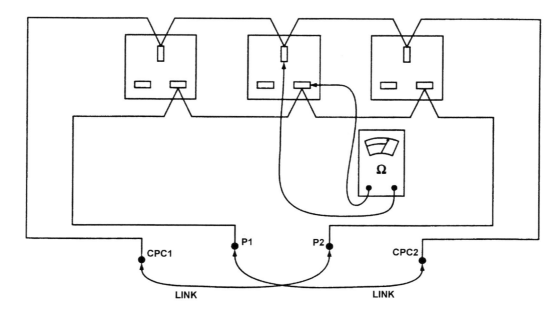

- Join the opposite ends of phase and cpc at the distribution board and measure between phase and cpc at every outlet or point on the circuit.

- The values obtained should be substantially the same – approximately ¼ of the phase ring plus ¼ of the cpc ring. Any sockets wired as spurs will have a higher reading in proportion to the length of spur.

- Record the results – the phase-phase (end-to-end) measured value is known as r_1, the neutral-neutral is r_n and cpc-cpc is r_2. Usually, the maximum value of R1 + R2 is the only one recorded on the Schedule of Test Results.

- Remember to reconnect all the conductors.

Note: When testing at the outlets for Step 2 or 3, if the readings at the outlets increase towards the centre of the ring and decrease back towards the distribution board, then it is likely that the opposite ends of the ring conductors have not been crossed as intended.

Where the cpc is formed by metal conduit or trunking, this metalwork must still be tested for continuity. However, circuit polarity will need to be confirmed separately.

Example

A 60 meter end-to-end ring circuit wired in 2.5 mm² conductors with 1.5 mm² cpc. The circuit has no spurs.

The results of the test will be approximately as follows:

Step 1 Phase-phase (r_1) reading	=	0.45 Ω
Neutral-neutral (r_n) reading	=	0.45 Ω
Cpc-cpc (r_2) reading	=	0.72 Ω
Step 2 Phase-neutral reading at outlets	=	0.23 Ω
Step 3 Phase-cpc ($R_1 + R_2$) reading at outlets	=	0.29 Ω

It can therefore be seen that, for socket outlets wired on a true ring circuit, the readings at the socket outlets in Step 2 are approximately ½ the readings in Step 1 (i.e. ¼ phase + ¼ neutral) and the values of Step 3 are approximately a ¼ of the phase + ¼ of the cpc values of Step 1.

If this test is carried out correctly, the results are satisfactory and the circuit has a separate cpc, then polarity will also be confirmed at each outlet. Also, the need for a separate cpc continuity test will be removed.

Insulation Resistance

These tests are to verify that the insulation of conductors, electrical accessories and equipment is satisfactory and that there is no unwanted path between live conductors or live conductors and earth.

Test instrument

An insulation resistance tester should be used which is capable of providing a DC voltage of not less than twice the nominal voltage of the circuit to be tested (rms value for an AC supply). The instrument should provide the following test voltages of Table 71A when loaded with 1 mA.

	Circuit nominal voltage	Test voltages	Minimum value
1	SELV and PELV	250 V DC	0.25 MΩ
2	All circuits (other than SELV or PELV) up to and including 500 V	500 V DC	0.5 MΩ
3	In excess of 500 V	1000 V DC	1.0 MΩ

Test procedure

- Safely isolate the supply, including neutral

- Ensure that all current-using equipment is disconnected (including neons, capacitors, discharge lighting, etc.) and all filament lamps are removed. Where it is impracticable to disconnect or remove current-consuming equipment, the local switches controlling this equipment should be left open. This will avoid any damage to equipment and prevent misleading results being obtained due to component resistances

- Disconnect control equipment or apparatus constructed with voltage sensitive devices (semiconductors) e.g. dimmer switches, timers, touch switches, electronic control gear for lighting, and will include certain types of RCD. The devices will be liable to damage if exposed to the high test voltages used in insulation resistance tests

  Apart from disconnecting the above, the installation should be as complete as possible with accessories fitted

- Ensure switches and circuit breakers are closed, and all fuses are in place

- Prepare the test instrument for use, checking that the leads are in good order and the instrument is working correctly on the voltage range for the circuit or installation under test

- Carry out the tests at the distribution board/consumer unit as required for the installation or circuit under test and record the readings

Note: If a reading of less than 2 MΩ is obtained then the reason for this low reading should be investigated.

Insulation Resistance Tests Between Live Conductors

Single-phase installations

Test:

- phase to neutral (any two-way switches to be operated)

Record the lowest reading in the test result schedule.

Insulation Resistance Tests to Earth

Where the circuit includes electronic devices (that are impracticable to disconnect), a measurement to protective earth can be made with the phase and neutral joined together. Precautions may be required to prevent damage to these electronic devices (713-04-04).

Single-phase installations

Test:

- Phase and neutral (connected together) to earth
  (any two-way switches to be operated)

*Note: Where circuits or equipment are not sensitive to insulation tests, phases and neutrals may be tested **separately** to earth.*

Equipment

When fixed equipment such as cookers have been disconnected to allow insulation resistance tests to be carried out, the equipment itself must be insulation-resistance tested between live points and exposed conductive parts.

The test results should comply with the appropriate British Standards. If none are applicable, the insulation resistance should not be less than 0.5 MΩ.

Electrical Separation of Circuits

Great care is needed when inspecting and testing separated extra low voltage (SELV), protective extra low voltage (PELV) or electrically separate circuits. An inspection of the installation should be made to verify the separation of the separated circuits and a range of insulation tests (the values in accordance with Table 71A) carried out, to ensure that electrical separation has been achieved (see IEE Guidance Note No. 3).

Protection by Barriers or Enclosures

Where barriers and enclosures have provided during erection (as 412-03) or if a barrier or enclosure of factory-built equipment has been affected during installation, it must be confirmed that the degree of protection of IP2X, IPXXB, or IP4X (as applicable) has been achieved (see IEE Guidance Note No. 3).

IP2X or IPXXB is the required degree of protection to prevent direct contact. For accessible top surfaces of an enclosure, IP4X is required to prevent entry of foreign objects greater than 1 mm diameter or width.

Polarity

This test must be carried out to verify that:

- all fuses, circuit-breakers and single pole control devices such as switches are connected in the phase conductor only

- except for E14 and E27 lampholders to BS EN 60238, all centre contact bayonet or Edison screw lampholders should have their outer screwed contact connected to neutral

- socket outlets and similar accessories have been correctly installed

The installation must be tested with all switches in the 'on' position and all lamps and power consuming equipment removed.

Test instrument

Low-resistance ohmmeter

Test procedure

- Note that this test is the same as the cpc continuity test method 1 – the ($R_1 + R_2$) version

- If correct polarity is verified using this test, a tick is usually placed in the polarity column of the test results schedule

A test of polarity can be carried out as illustrated below.

Polarity test ($R_1 + R_2$ method)

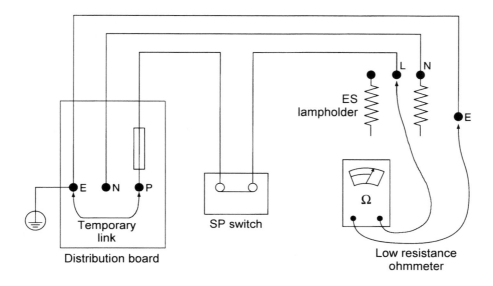

Polarity test: lighting

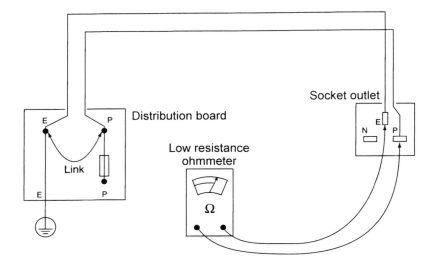

Polarity test: socket outlet

For ring final circuits that contain a separate cpc, if the circuit has been tested to 713-03 then polarity will have been confirmed at each outlet during the test.

After connection of the supply, polarity should be reconfirmed with a voltage indicator. This will establish the correct reconnection of any conductors previously disconnected during the testing process.

Phase Earth Fault Loop Impedance

The earth fault current loop comprises the following parts:

- the point of fault
- circuit protective conductor
- the main earthing terminal
- earthing conductor
- for TN systems, the metallic return path – or, with TT systems, the earth return path (through the mass of earth)
- the earthed neutral point of the transformer
- the transformer winding
- the phase conductor from the transformer to the point of fault

The earth fault loop path of a TN-S system is illustrated below.

Note: The impedance of the earth fault loop is denoted by the symbol Zs
Where Z = impedance, S = system (a supply of energy and an installation)

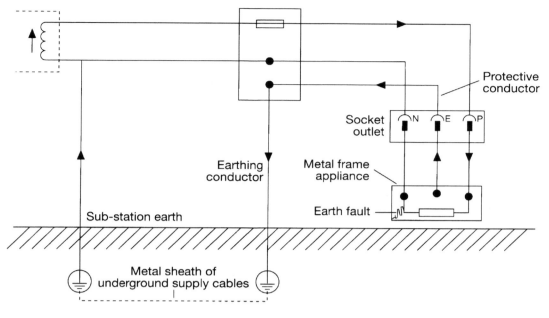

General mass of earth

Test instrument

An earth loop impedance tester.

The earth loop impedance tester uses the circuit voltage to pass a test current of around 20 amperes through the phase-earth path for a duration of approximately 20-30 milli-seconds. The current passes through a known resistor within the instrument and the instrument compares the circuit when loaded and not loaded with this resistance and displays the impedance value in ohms.

When using this test instrument, care must be taken to ensure that no ill effects can arise in the event of any defect in the earthing circuit, such as would arise if there was a break in the protective conductor of the system under the test. This would prevent the test current from flowing and the whole of the protective conductor system would be connected directly to the phase conductor.

Important: Live polarity must be re-confirmed before carrying out this test.

Test procedure

- This is a live test so be aware of the safety precautions to prevent danger

- Prepare the instrument for use, checking that the instrument and leads, probes and clips are in good order and suitable for purpose

- If using the probes or clips, care should be taken to minimise the risk of electric shock or burns due to working on or near exposed live terminals. Always connect the instrument securely before turning on the supply; then switch off the supply after carrying out the test and before disconnecting the instrument

- Before carrying out a test, check the status of the LEDs or neons on the instrument to see if they indicate it is safe to proceed with the test. If they indicate problems with the wiring then, for safety reasons, these problems must be rectified before carrying out the test

- With socket outlets, test the furthest one on each circuit as a minimum. Otherwise, test all the socket outlets in the circuit and record the highest reading (testing at socket outlets requires the instrument manufacturer's lead with fitted plug)

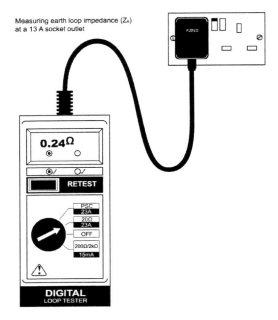

Measuring earth loop impedance (Z_s) at a 13 A socket outlet

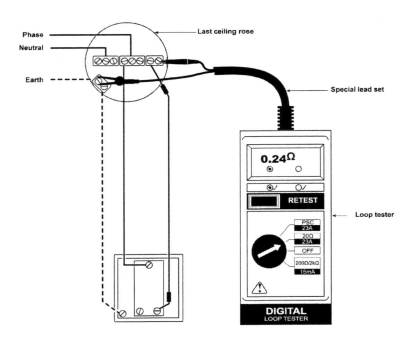

- When testing lighting circuits, the test (as a minimum) should be carried out at the electrically furthest point of the circuit. This could be at a luminaire or switch

- Earth fault loop impedance (Zs) should be verified at the furthest point of each circuit, e.g. distribution lighting and socket outlet circuits and any fixed equipment
- The test result value of Zs should be checked for compliance by being compared to:
  - the values in Appendix 2 of the IEE On-Site Guide or IEE Guidance Note No. 3 (for standard PVC circuits), OR
  - values in BS 7671 when corrected for temperature, OR
  - rule of thumb method

  Each of these methods is described in the following sections.

IEE On-Site Guide, Appendix 2: maximum measured earth loop impedance

The values of maximum earth loop impedance in Tables 2A–D of the On-Site Guide and IEE Guidance Note No. 3 are maximum **measured** values at a testing temperature of 10°–20°C, but these will need correcting if the ambient temperature at the time of test falls outside this range (using the correction factors from Table 2E).

For example, if the ambient temperature is 25°C then the maximum measured phase earth loop impedance for a socket outlet circuit protected by a 32 A Type B circuit breaker would be:

$$1.20 \times 1.06 = \mathbf{1.27 \; \Omega}$$

BS 7671 values, corrected for temperature

The IEE Wiring Regulations give maximum values of phase earth loop impedance to achieve maximum disconnection times should an earth fault occur.

Measured values of phase earth loop impedance should subsequently be less than those in BS 7671 Tables 41B1 or 41B2 and 41D. The values found in these tables can be corrected by multiplying by a correction factor for ambient temperature (at the time of testing) and by a correction factor for the conductor temperature.

The value of Ze must first be deducted from the measured value of Z_s. Only the measured values of $R_1 + R_2$ are corrected as the value of Z_e is unlikely to be affected by temperature. However, it must be added back on to the corrected values of measured $R_1 + R_2$ to give a corrected value of Z_s.

This corrected value must then be compared to the values of the appropriate table – BS 7671 Tables 41B1 or 41B 2, 41D; 604B1 or 604B2; or 605B1 or 605B2.

Example

See Tables 1B, 1C and 1D of Guidance Note No. 3 or Tables 9B, 9C and 9D of the IEE On-Site Guide.

For a given circuit:

- maximum tabulated value of Z_s is **3.43 Ω**
- Measured value of Z_s is **2.5 Ω**
- The value of Z_e is **0.3 Ω**

If the ambient temperature at the time of test was 15°C, **factor 0.98** (Table 1B/9B)

The protective device for the circuit was a standard device from Table 41B1 and the protective conductor was a core in a cable that had 70°C PVC insulation, factor 1.2 (Table 1C/9C).

- $Z_s - Z_e = R_1 + R_2$ value

    2.5 – 0.3 = 2.2

- $R_1 + R_2$ x factors applicable

    2.2 x 0.98 x 1.2 = 2.58

- Add corrected $R_1 + R_2$ to Z_e

    2.58 + 0.3 = 2.88 Ω

- **Corrected Z_s = 2.88 Ω**

- This value should now be compared to the maximum value of Z_s in the appropriate column of Tables 41B or 41D as applicable

    This corrected value should be equal to or less than the maximum value that is stated in tables.

    Corrected Z_s = 2.88 Ω

    Maximum tabulated value of Z_s = 3.43 Ω

    Since 2.88 < 3.43, the result is satisfactory

Rule of thumb method

Alternatively, as a general guide, the measured value should not exceed 80% of the values stated in Tables 41B1 or 41B2, 41D; 604B1 or 604B2; 605B1 or 605B2 as appropriate. (Refer to maximum earth fault loop impedance – Appendix A)

If test values are found to be close to this limit, the circuit should be inspected to find the reason, for example, loose connections, etc.

External Earth Fault Loop Impedance (Ze)

Z_e is the earth fault loop impedance of the supply, i.e. that part of the system which is external to the installation.

Since the purpose of this test is to establish that the means of earthing is present and it is of an acceptable ohmic value, any parallel paths must be disconnected. This requires the disconnection of the earthing conductor, otherwise misleading readings would be obtained and defects, or even the lack of earthing, could be concealed.

For safety reasons, the whole installation must be safely isolated from the supply before the earthing conductor is disconnected and must stay isolated while the test is carried out and until the earthing conductor is reconnected correctly. Only then can the supply be restored.

Test instrument

An earth loop impedance tester.

Some earth loop impedance testers have two leads, while others have three leads. With the two-lead set, the connection is to the incoming phase and earth; with the three-lead set, a neutral connection is required. Where the installation under test is three-phase and no neutral is present, normally the neutral lead of the instrument is connected to earth. Always refer to the instrument manufacturer's instructions for the correct connections.

Test procedure

- Safely isolate the whole installation from the supply
- Disconnect the earthing conductor
- Prepare the instrument for use, checking the instrument and leads, probes and clips are in good order and suitable for purpose
- Remember to take great care during the test as it is at the origin where the highest levels of fault conditions can be found
- Following the safety procedures, connect one lead of the instrument to the earthing conductor then, using a probe, connect the other lead to the incoming live phase conductor (checking any instrument polarity indication)
- Carry out the test and record the reading obtained
- **Reconnect the earthing conductor correctly**
- Restore the supply

Note: For three-phase installations, a test must be carried out separately on each supply phase conductor.

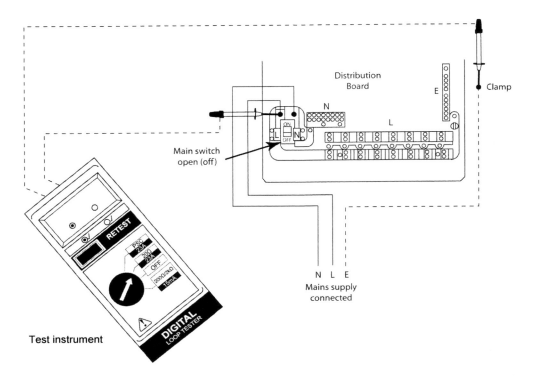

The reasons for carrying out the test are to verify there is an earth connection and to establish that the actual Z_e value is equal to or less than the value used by the electrical designer in his/her calculations.

Values for Z_e can also be obtained by enquiry to the local distributor or by calculation. These are the only two possible methods of establishing Z_e if the installation cannot be safely isolated.

Sensitive protective devices

Carrying out an earth fault loop impedance test will cause sensitive protective devices (such as RCDs, and possibly 6 amp Type B circuit breakers and any 6 amp Type 1 mcbs) of existing installations to operate. In these circumstances, the following options are available.

- Instrument manufacturers can supply instruments that:
  - limit the test current (typically to 15 mA, as this means an RCD with a 30 mA or greater operating current should not trip), OR
  - DC bias the RCD (this method momentarily stuns the RCD by saturating the core of the device so the test current is not detected)

- An earth loop test can be carried out on the incoming supply to the RCD with the circuit's R1 + R2 value being added to the test result to give an approximate value of Zs.

Testing an RCD-protected socket outlet by limiting the test current is shown below – note the 15 mA setting.

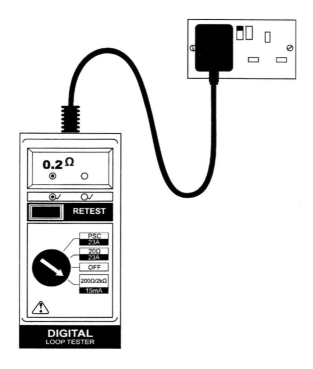

Prospective Fault Current *(713-12)*

Both the prospective short circuit current and the prospective earth fault current (PFC) should be measured, calculated or determined by enquiry at the origin and other relevant points in the installation, typically other distribution boards throughout the installation (434-02-01).

The value of PFC must be determined to establish that the type of protective device at the origin would interrupt a fault current up to, and including, the level of PFC that could flow if a short circuit or earth fault occurs.

The short circuit (duty) rating of any protective device must be equal to or greater than the value of PFC that could occur at the point where the device is installed. The exception to this rule is where a local device with a lower duty rating is backed up by a device with a higher duty rating.

An example would be a domestic installation with a 100 A supply and 16KA PFC value, with a 100 A 80KA BS 88 service fuse in the distributor's cut-out, and circuit breakers with a 6KA duty rating in the installation consumer unit. Therefore, this value of PFC (prospective fault current covers both prospective short circuit current and prospective earth fault current) must be determined and compared to the short circuit duty rating of the relevant protective device.

Test instrument

An earth loop impedance/PFC tester

Some earth loop impedance/PFC testers use three leads, others use only two leads. Always refer to the instrument manufacturer's instructions for the correct connections.

Test procedure

- Great care is required during the testing procedure. Fault currents are at their highest values at the origin of an installation where this live test will be carried out

- Prepare the instrument for use, checking that the instrument and leads, probes and clips are in good order and suitable for purpose

- Select the correct function and range on the instrument (KA)

- The tests should be conducted at the main switch or wherever the tails from the distributor's metering equipment are connected

- Being aware of all the safety requirements, connect the instrument (in accordance with the manufacturer's instructions) to the incoming live supply to measure the phase to neutral value of PFC. See diagram A below

- Observe the polarity indication on the instrument for correct connection

- Carry out the test and record the readings

- Now repeat the process, but with the instrument being connected (in accordance with the manufacturer's instructions) to measure the phase to earth value of PFC. See diagram B on the following page

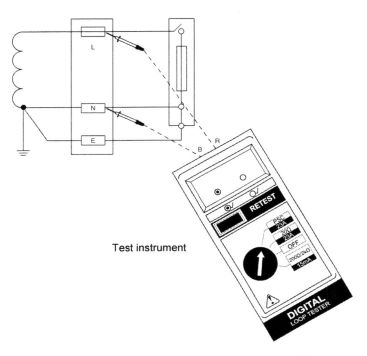

Test instrument

**A: Measurement of prospective short circuit current
(using a two-lead instrument)**

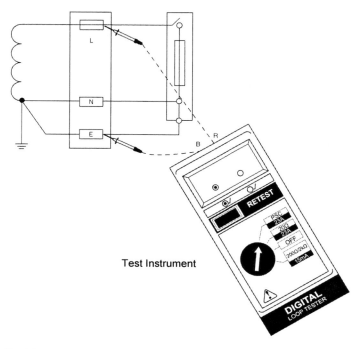

**B: Measurement of prospective earth fault current
(using a two-lead instrument)**

The result to be recorded is the highest of the phase to neutral or phase to earth value.

Note: For three-phase installations, the highest value of PFC will be phase to phase.
Do not attempt this test with a 230 volt instrument.

The value can be found by calculation; it will be approximately double the phase to neutral fault current.

Fault current protective devices installed at the origin of an installation must have a short circuit duty rating equal to or greater than the level of fault current that would flow at their point of installation.

Functional Testing

RCDs manufactured to BS 4293 or BS EN 61008, RCBOs to BS EN 61009 and RCD socket outlets (SRCDs) should be tested by simulating appropriate fault conditions, using a test instrument.

The test is made on the load side of the RCD, between the phase conductor of the circuit protected and its associated circuit protective conductor. All loads normally supplied through the RCD are disconnected during the test.

The test should not be carried out until it has been established that the circuit earth loop impedance value is sufficiently low to allow the testing of the RCD (as required by BS 7671) i.e. the impedance value multiplied by the RCD operating current must not exceed 50 V (or 25 V for a construction site or agricultural installation).

Test instrument

RCD tester

This is a live test so it is important to check the status of the instrument LEDs or neons to see if they indicate it is safe to carry out the tests.

The test is carried out on the load side of the RCD (with the load disconnected). The connection can be made at any suitable point in the circuit, e.g. a socket outlet or at the output terminals of the RCD using the probe lead. However, make sure you have correct polarity by checking the correct polarity indication LEDs or neons on the instrument.

Test procedure

Testing a 30 mA IΔn RCD-protected socket outlet:

- plug in the RCD tester and check that the polarity indication on the instrument shows it is safe to carry out the test

- if it is safe to proceed, carry out the three instrument tests required

- with a fault current of 50% of the trip current of the RCD flowing for a period of 2 secs, the RCD should **not** open (sometimes called the x½ test)

- with 100% of the rated tripping current flowing, the circuit breaker must open within 200 ms (x1 test)

- supplementary protection
  When the RCD has an operating current IΔn not exceeding 30 mA and has been installed to reduce the risk associated with direct contact as indicated in Regulation 412-06-02, a test current of 5 x IΔn should cause the circuit breaker to open within 40 ms, which is the maximum test time (x5 test)

- the tests should be carried out on both the positive and negative half cycles of the supply and the longest operating time should be recorded

The effectiveness of the integral test facility of the RCD should also be verified by pressing the test button. Users are advised to carry out this test quarterly.

Note: This should be carried out after the instrument tests so as not to affect their results.

If users have not operated the RCD by means of the test button and depending also on the environmental conditions where the RCD is installed, the device may not operate on the first test (50%) or for the 100% test. After manual operation, however, the device may trip as required and, for this reason, some test engineers would carry out the 50% test after the other instrument tests have been carried out.

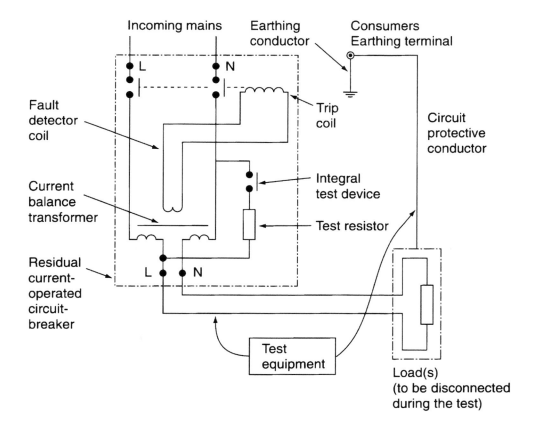

There are also additional test requirements for RCDs that incorporate time delays, etc. See IEE Guidance Note No. 3 for further information.

Alterations and Additions to an Installation

The requirements of Chapter 71 (initial verification) are applicable to alterations and additions.

Every addition or alteration should be in compliance with the regulations and should not impair the safety of an existing installation.

The maximum period between inspections is:

Domestic installations   10 years

Certification and Reporting

General

On completion of the verification of a new installation or alterations to an existing one, an Electrical Installation Certificate should be provided. It should include details of the extent of the installation covered, along with schedule(s) of inspection, and schedule(s) of test results.

A Periodic Inspection Report that includes details of the extent of the installation and any limitations of the inspection and testing, along with schedule(s) of inspection and schedule(s) of test results should be provided for a periodic inspection and test (741-01-02).

If minor installation works does not include a new circuit, but relates to an addition or alteration to an existing circuit, a Minor Electrical Installation Works Certificate should be provided.

An Electrical Installation Certificate, Minor Electrical Installation Works Certificate or Periodic Inspection Report should be signed by a competent person(s).

Initial Verification

After the completion of an initial verification, an Electrical Installation Certificate with schedule(s) of inspection and schedule(s) of test results should be given to the person who ordered the work.

The identification of every circuit, including its protective devices, and the record of relevant test results should appear on the schedule(s) of test results.

Any defects or omissions should be made good before the certificate is issued.

Alterations and Additions

Sections 741 and 742 apply to all the work of an alteration or addition. Any defects found in the existing installation should be recorded in the Electrical Installation Certificate or Minor Electrical Installation Work Certificate.

Periodic Inspection and Testing

A Periodic Inspection Record, together with schedule(s) of inspection and schedule(s) of test results, should on completion of the inspection and testing be given to the person who ordered the inspection.

Any deterioration, damage defects, dangerous conditions and non-compliances which may present a risk of danger should be recorded along with the extent and limitations of the inspection.

Any condition that presents an immediate danger should be rectified, otherwise the defect should be reported in writing to the responsible person.

Test Instruments

Test instruments should be regularly checked and re-calibrated to ensure accuracy, usually on an annual basis. The serial number of the instrument used should be recorded with test results, to avoid unnecessary re-testing if on of a number of instruments is found to be inaccurate.

For operation, use and care of test instruments, refer to the manufacturer's handbook.

Note: Attention is drawn to HSE Guidance Note GS38 "Electrical test equipment for use by electricians' published by HMSO, which advises on the selection and safe use of suitable test probes leads, lamps, voltage indicating devices and other measuring equipment.

APPENDICES

APPENDIX A

National Association of
Professional Inspectors & Testers

MAXIMUM EARTH FAULT LOOP IMPEDANCE
as Tables 41B1, 41B2 and 41D

P/Device Rating	PROTECTIVE DEVICE TYPE & BS NUMBER													
	BS:88 PART 2 & 6		BS:1361		BS:3036		BS:1362		BS:3871 TYPE			BS EN 60898 TYPE		
									1	2	3	B	C	D
	0.4s	5s	0.4s	5s	0.4s	5s	0.4s	5s	0.4/5s	0.4/5s	0.4/5s	0.4/5s	0.4/5s	0.4/5s
5	N/A	N/A	10.9	17.1	10	18.5	N/A	N/A	12	6.86	4.8	N/A	4.8	2.4
6	8.89	14.1	N/A	N/A	N/A	N/A	N/A	N/A	10	5.71	4	8	4	2
10	5.33	7.74	N/A	N/A	N/A	N/A	N/A	N/A	6	3.43	2.4	4.8	2.4	1.2
13	N/A	N/A	N/A	N/A	N/A	N/A	2.53	4	N/A	N/A	N/A	N/A	N/A	N/A
15	N/A	N/A	3.43	5.22	2.67	5.58	N/A	N/A	4	2.29	1.6	N/A	1.6	0.8
16	2.82	4.36	N/A	N/A	N/A	N/A	N/A	N/A	3.75	2.14	1.5	3	1.5	0.75
20	1.85	3.04	1.78	2.93	1.85	4	N/A	N/A	3	1.71	1.2	2.4	1.2	0.6
25	1.5	2.4	N/A	N/A	N/A	N/A	N/A	N/A	2.4	1.37	0.96	1.92	0.96	0.48
30	N/A	N/A	1.2	1.92	1.14	2.76	N/A	N/A	2	1.14	0.8	N/A	0.8	0.4
32	1.09	1.92	N/A	N/A	N/A	N/A	N/A	N/A	1.88	1.07	0.75	1.5	0.75	0.38
40	0.86	1.41	N/A	N/A	N/A	N/A	N/A	N/A	1.5	0.86	0.6	1.2	0.6	0.3
45	N/A	N/A	0.6	1	0.62	1.66	N/A	N/A	1.33	0.76	0.53	1.07	0.53	0.27
50	0.63	1.09	N/A	N/A	N/A	N/A	N/A	N/A	1.2	0.69	0.48	0.96	0.48	0.24
60	N/A	N/A	N/A	0.73	N/A	1.17	N/A	N/A	N/A	N/A	N/A	N/A	N/A	N/A
63	N/A	0.86	N/A	N/A	N/A	N/A	N/A	N/A	0.95	0.54	0.38	0.76	0.38	0.19
80	N/A	0.6	N/A	0.52	N/A	N/A	N/A	N/A	0.75	N/A	N/A	N/A	N/A	N/A
100	N/A	0.44	N/A	0.38	N/A	0.56	N/A	N/A	0.60	N/A	N/A	N/A	N/A	N/A
125	N/A	0.35	N/A	N/A	N/A	N/A	N/A	N/A	0.48	N/A	N/A	N/A	N/A	N/A
160	N/A	0.27	N/A	N/A	N/A	N/A	N/A	N/A	0.37	N/A	N/A	N/A	N/A	N/A
200	N/A	0.2	N/A	N/A	N/A	N/A	N/A	N/A	0.30	N/A	N/A	N/A	N/A	N/A
250	N/A	0.16	N/A	0.52	N/A	N/A	N/A	N/A	0.24	N/A	N/A	N/A	N/A	N/A
315	N/A	0.12	N/A	0.38	N/A	0.56	N/A	N/A	N/A	N/A	N/A	N/A	N/A	N/A
400	N/A	0.09	N/A	N/A	N/A	N/A	N/A	N/A	N/A	N/A	N/A	N/A	N/A	N/A
500	N/A	0.065	N/A	N/A	N/A	N/A	N/A	N/A	N/A	N/A	N/A	N/A	N/A	N/A

APPENDIX A

National Association of
Professional Inspectors & Testers

MAXIMUM EARTH FAULT LOOP IMPEDANCE at 80% of Tables 41B1, 41B2 and 41D

P/Device Rating	BS:88 PART 2 & 6		BS:1361		BS:3036		BS:1362		BS:3871 TYPE			BS EN 60898 TYPE		
									1	2	3	B	C	D
	0.4s	5s	0.4s	5s	0.4s	5s	0.4s	5s	0.4/5s	0.4/5s	0.4/5s	0.4/5s	0.4/5s	0.4/5s
5	N/A	N/A	8.72	13.7	8	14.8	N/A	N/A	9.6	5.5	3.84	N/A	3.84	1.92
6	7.11	11.3	N/A	N/A	N/A	N/A	N/A	N/A	8	4.57	3.2	6.4	3.2	1.6
10	4.26	6.19	N/A	N/A	N/A	N/A	N/A	N/A	4.8	2.74	1.92	3.84	1.92	0.96
13	N/A	N/A	N/A	N/A	N/A	N/A	2.02	3.2	N/A	N/A	N/A	N/A	N/A	N/A
15	N/A	N/A	2.74	4.17	2.14	4.46	N/A	N/A	3.2	1.83	1.28	N/A	1.28	0.64
16	2.25	3.48	N/A	N/A	N/A	N/A	N/A	N/A	3	1.71	1.2	2.4	1.2	0.6
20	1.48	2.43	1.42	2.34	1.48	3.2	N/A	N/A	2.4	1.37	0.96	1.92	0.96	0.48
25	1.2	1.92	N/A	N/A	N/A	N/A	N/A	N/A	1.92	1.09	0.77	1.53	0.77	0.38
30	N/A	N/A	0.96	1.53	0.91	2.2	N/A	N/A	1.6	0.91	0.64	N/A	0.64	0.32
32	0.87	1.53	N/A	N/A	N/A	N/A	N/A	N/A	1.20	0.85	0.6	1.2	0.6	0.30
40	0.68	3.52	N/A	N/A	N/A	N/A	N/A	N/A	1.2	0.68	0.48	0.96	0.48	0.24
45	N/A	N/A	0.48	0.8	0.49	1.32	N/A	N/A	1.06	0.60	0.42	0.85	0.42	0.21
50	0.50	0.87	N/A	N/A	N/A	N/A	N/A	N/A	0.96	0.55	0.38	0.76	0.38	0.19
60	N/A	N/A	N/A	0.58	N/A	0.93	N/A	N/A	N/A	N/A	N/A	N/A	N/A	N/A
63	N/A	0.68	N/A	N/A	N/A	N/A	N/A	N/A	0.76	0.43	0.30	0.60	0.30	0.15
80	N/A	0.48	N/A	0.41	N/A	N/A	N/A	N/A	N/A	N/A	N/A	N/A	N/A	N/A
100	N/A	0.35	N/A	0.30	N/A	0.45	N/A	N/A	N/A	N/A	N/A	N/A	N/A	N/A
125	N/A	0.28	N/A	N/A	N/A	N/A	N/A	N/A	N/A	N/A	N/A	N/A	N/A	N/A
160	N/A	0.21	N/A	N/A	N/A	N/A	N/A	N/A	N/A	N/A	N/A	N/A	N/A	N/A
200	N/A	0.16	N/A	N/A	N/A	N/A	N/A	N/A	N/A	N/A	N/A	N/A	N/A	N/A
250	N/A	0.16	N/A	0.52	N/A	N/A	N/A	N/A	N/A	N/A	N/A	N/A	N/A	N/A
315	N/A	0.12	N/A	0.38	N/A	0.56	N/A	N/A	N/A	N/A	N/A	N/A	N/A	N/A
400	N/A	0.09	N/A	N/A	N/A	N/A	N/A	N/A	N/A	N/A	N/A	N/A	N/A	N/A
500	N/A	0.065	N/A	N/A	N/A	N/A	N/A	N/A	N/A	N/A	N/A	N/A	N/A	N/A

APPENDIX B

NAPIT Electrical Certificate (Minor Works)

For compliance with Building Regulations Part P

NOTES:
1. This Minor Works Electrical Certificate **shall only be used** for the reporting on the condition of an electrical installation, where the work does not comprise the addition of a new circuit.
2. This Certificate is based upon the format of the Minor Electrical Installation Works Certificate, issued by the Institute of Electrical Engineers and published in BS7671:2001
3. The Inspection, Test, Verification of compliance to BS7671:2001 and Completion of this Certificate must be undertaken by a person on a Competent Persons Scheme approved by the Secretary of State or on the NAPIT Register of Approved Electrical Inspectors.

NAPIT Electrical Certificate (Minor Works)

Information for recipients (to be appended to the report).

This Minor Works Electrical Certificate **shall only be used** for the reporting on the condition of an electrical by installation, where the work does not comprise the addition of a new circuit.

You should have received an original Report and the contractor should have retained a duplicate.

If you were the person ordering this Report, but not the owner of the installation, you should pass this Report, or a copy of it, immediately to the owner.

The original Report is **to be retained in a safe place and be shown to any person inspecting or undertaking work on the electrical installation in the future.**

If you later vacate the property, this Report will provide the new owner with details of the condition of the minor electrical installation works at the time the Report was issued.

For safety reasons, the electrical installation will need to be re-inspected at appropriate intervals by a competent person.

The maximum time interval recommended before an inspection of the installation is stated in the Report under 'An Inspection' [5(4)] and the Institute of Electrical Engineers recommend every 10 years or on change of occupancy for Domestic Electrical Installations

If this work is carried out by a person on a Competent Persons Register approved by the Secretary of State, you should also receive a 'Compliance with Building Regulations Declaration' **within 30 days of the electrical installation being completed,** if not then this work may be subject to Local Authority Building Regulation certification.

Caution:

An initial assessment of the existing electrical installation must be made to ensure that the proposed Minor Electrical Installation Works can be undertaken safely, this assessment must be undertaken by a person competent to do so.

Part 1 : Initial Assessment of the existing circuit to be extended

BS Number and Type of Circuit Protection Device BSEN60898B Current Rating 32

Number of existing accessories already connected to this circuit 4 Sockets Circuit number 2

1. Is the addition to the circuit in an area of high risk? (Yes) No

   if yes, is it a (Kitchen) / Bathroom / Outdoors / if Other please state _____

2. System Earthing arrangement: TN-C-S ☑   TN-S ☐   TT ☐

3. Method of protection against indirect contact: (EEBAD) / if Other please state _____

4. Is there enough spare load capacity on the existing system to allow it to be extended? (Yes) No

5. Have Tests been carried out prior to installation works to ensure the circuit to be extended is healthy (Yes) / No

5. Are the existing Earthing and Bonding arrangements adequate? (Yes) No

Comments on the Initial Assessment (if none write 'none' below)

NONE

Copyright NAPIT @ November 2004

1 of 2

APPENDIX B

NAPIT Electrical Certificate (Minor Works)

NAPIT Electrical Certificate (Minor Works) (note 1)

Only to be used for minor electrical work, this does not include the provision of a new circuit.

Part 2 : Description of the minor works

1. Description of the minor works : FUSED SPUR   Located: KITCHEN. ADDED TO EXISTING RING CIRCUIT
2. Address of the property: 5, LAWMUIR PLACE, MOTHERWELL
4. Date the minor works completed : 14TH APRIL 2005

Part 3 : Visual Check at 'First Fix' Stage (1st Column) Inspection Results (2nd Column) (tick, cross, N/C or N/A)

Check	1st	2nd	Check	1st	2nd
Main Equipotential Bonding present and adequate	✓	✓	All Cable cores correctly identified in joints and accessories	✓	✓
Correct Circuit Protection Device fitted and identified	✓	✓	Appropriate Supplementary Bonding present and adequate	✓	✓
Correct Cable type and size used, allowing for external influences	✓	✓	Modifications to the Building Fabric appropriate and safe	✓	✓
Cable run in 'safe' zones or mechanically protected	✓	N/C	All Accessaries correctly placed as per Approved Document M and BS 8300	✓	✓
Cables securely fastened or in appropriate wiring protection systems	✓	✓	All covers replaced, Accessories secure and neatly aligned.	N/C	✓
All cable joints correctly terminated, secure and accessable	✓	✓	Circuit details updated on the installation schedule (or schedule produced)	N/A	✓

Name: JOHN GLEN   Signed: John Glen   Position: INSPECTOR   Date: 14/04/05

Part 4 : Essential Tests

1. Earth Continuity: Value (R1+R2 or R2) 0.27 Ω. If R2 is this Satisfactory (table 41C)? **Yes** / No

2. Insulation resistance:
   - Phase/Neutral >2000 MΩ
   - Phase/Earth >2000 MΩ
   - Neutral/Earth >2000 MΩ

3. Polarity : Satisfactory **Yes** / No

4. Earth fault loop impedance 0.54 Ω

5. RCD (if applicable) : Rated residual operating current $I_{\Delta n}$ 30 mA and operating time of 32 ms (at $I_{\Delta n}$) 14 ms (at 5$I_{\Delta n}$)

PART 5 : Declaration

1. I/We CERTIFY that the said works do not impair the safety of the existing installation, that the said works, as far as it is reasonably practical to determine, have been designed, constructed, inspected and tested in accordance with BS 7671:2001 (IEE Wiring Regulations), amended to 2004 and that the said works, to the best of my/our knowledge and belief, at the time of my/our inspection, complied with Chapter 13 of BS 7671:2001

2. Electrical Inspector: JOHN GLEN
   For and on behalf of: APOLO TESTING
   Address: 14 LUNAR TER. QUEENS WAY LONDON

3. Signature: John Glen
   Position: Inspector
   Date: 14/04/05

4. An Inspection of this installation is recommended after 10 years OR CHANGE OF OCCUPANCY

NAPIT Membership Number: 24687   Expiry Date: 17/12/06

2 OF 2

Initial Application Form

Trading Title:			**Standard Membership**
Contact Name:	Position:		**Initial Assessment Fee - £180.00**
Address:			This covers the cost of a company assessment and up to 2 individual assessments. For each additional inspector requiring an assessment, a fee of £50.00 will be charged. *The assessment fee is payable at the first stage application and is non-refundable.*
Town:	County:		
Post Code:	Email:		
Tel No:	Mobile No:		
Fax No:	Web address:		
Date Established:	VAT No:		
Number of Employees:	Company Reg No:*(if applicable)*		**Annual Membership** £300.00 p.a.
Electricians:	Engineers:		**Building Control notifications** - £1.50 + vat per on-line notification or £3.00 + vat per fax or post notification, plus £0.70 + vat per notification for the insurance backed guarantee.
Apprentices:	Technical Sub-Contract Staff (brief details):		

Level of Entry to the Competent Persons Register (if Required):

Level A	Level B	Just 8	
Fully Qualified Electrician	Other trades	Occasional level A (fully qualified Electrical Engineer only)	***Just8 Membership*** Must be in full time employment as a Professional Engineer, Consultant or Lecturer etc or semi-retired.

Details of individuals requiring approval: Please feel free to use a separate piece of paper for any additional persons (must be at the same time and on the same site during the assessment)

Name:	NI number:	**Initial Assessment Fee - £55.00** *The assessment fee is payable at the first stage application and in non-refundable*
Formal Qualifications held with dates gained:		**Annual Membership** £185.00 p.a.
		Building Control notifications – up to 8 notifications are included in the annual fee. More than 8 notifications will require full membership.
Name:	N.I number	**Please note:**
Formal Qualifications held with dates gained:		♦ Only those Electricians/Inspectors/Tradespersons listed will be registered and approved by the association. ♦ Only the Company Name/Branch/Department appearing on the application form will be registered with the association and carry the entitlement to display the association logo. ♦ Any further Company/Branch/Department registrations must be made on a separate application form and, if accepted, will be subject to an additional subscription
More than 2 x operatives requiring an assessment, attract an extra fee of £50 per person, if the assessment is carried out at the same time and venue as the company assessment. Any extra assessments required at a different venue or time, are at £180 per half day visit.		*Prices are accurate at the time of going to press.*

Please send a cheque or charge the card detailed below (delete as appropriate)

Card No:		Type (visa, Switch, etc)
Exp. Date		Issue No
Name on Card		Amount
Name: (Please print)		Position
Signed on behalf of:		Date:

NAPIT Administration Centre, 4th Floor, Mill 3, Pleasley Vale Business Park, Mansfield, Nottinghamshire, NG19 8RL
Tel: 0870 444 1392 Fax: 0870 444 14527

BES PUBLICATIONS

Building Engineering Services continue to provide the gas, electric, water and refrigerant industries with a range of popular, respected and competitively priced publications.

These publications can be used either as the basis of training or for reference in the workplace. Some can also be used for assessment purposes. All are published in A4 format, with the most popular also available as A5, pocket-sized books.

DOMESTIC GAS

GAS SAFETY (G1)
Format: A4 in a ringbinder

The complete manual for reference or self-study. All of the essentials in 300 pages, with clear explanations and illustrations, covering ◆gas pipework ◆gas supply ◆combustion ◆appliance gas safety devices and gas controls ◆principles of gas flues ◆flueing standards ◆ventilation requirements ◆emergency procedures ◆unsafe situations ◆warning notices and labels. Also included is the HSE publication ◆*Safety in the installation and use of gas systems and appliances* (G31) which covers the HSE Gas Safety (Installation and Use) Regulations 1998 – Approved Code of Practice and Guidance, a ◆*Course Workbook* and a booklet of ◆*Practical Tasks* for you to complete.

GAS SAFETY (G2)
Format: A5 Wiro-bound

All the information and diagrams from GAS SAFETY (G1) in a handy size for reference on the job and for carrying in the service van.

DOMESTIC GAS APPLIANCES (G5)
Format: A4 in a ringbinder

Contains all seven of the domestic natural gas appliance manuals from ConstructionSkills in one package, plus the *Domestic Natural Gas Appliances Course Workbook (G14)*. The easy-to-use format makes it ideal for engineers working with a range of domestic appliances.

Each manual can also be purchased individually:
- Heating Boilers/Water Heaters (G7)
- Cookers (G8)
- Ducted Air Heaters (G9)
- Fires and Wall Heaters (G10)
- Tumble Dryers (G11)
- Meters (G12)
- Instantaneous Water Heaters (G13)

DOMESTIC GAS APPLIANCES (G6)
Format: A5 Wiro-bound

All the information and diagrams from the DOMESTIC GAS APPLIANCES (G5) in a handy size for reference on the job and for carrying in the service van.

FAULT-FINDING TECHNIQUES (G17)
Format: A4

Problems with locating that elusive fault? Follow the step-by-step techniques in this hands-on manual and speed up your fault finding on central heating systems.

SAFETY IN THE INSTALLATION AND USE OF GAS SYSTEMS AND APPLIANCES (G31)
Format: A4

An essential HSE publication for all those working with domestic gas. It gives advice on how to comply with *The Gas Safety (Installation and Use) Regulations 1998 – Approved Code of Practice and Guidance,* which has a special legal status. For example, if you are prosecuted for breach of health and safety law, and it is proved that you have not followed the relevant parts of the Code, a court will find you at fault (unless you can show that you have complied with the law in some other way).

COMMERCIAL AND INDUSTRIAL GAS

COMMERCIAL GAS SAFETY (G88)
Format: A4 in a ringbinder

An essential training and reference manual for those working in the commercial environment. It includes key sections from the popular GAS SAFETY (G1) and incorporates information from two other commercial publications (G23 and G24) which can be purchased separately) making this the definitive training and reference manual for commercial work. It covers ◆commercial gas safety ◆pipework and ancillary equipment ◆gas pipework ◆gas supply ◆combustion ◆appliance gas safety devices and gas controls ◆principles of gas flues ◆flueing standards ◆ventilation requirements ◆emergency procedures ◆unsafe situations ◆warning notices and labels. Also included is the HSE publication ◆*Safety in the installation and use of gas systems and appliances* (G31) and ◆*Course Workbooks* and *Practical Tasks* (G3, G4, G83 and G84).

COMMERCIAL GAS SAFETY (G23)
Format: A4

An essential supplement for engineers working in the commercial environment. If you already own a GAS SAFETY (G1) pack, all you need is this book with its commercial gas-specific sections ◆combustion and flue gas analysis ◆burners ◆controls and control systems ◆flues ◆ventilation ◆pressure and flow.

COMMERCIAL PIPEWORK AND ANCILLARY EQUIPMENT (G24)
Format: A4

An essential guide for engineers working on commercial pipework, with clear information on ◆pipework design ◆soundness testing and purging ◆commercial metering ◆boosters and compressors.

COMMERCIAL APPLIANCES (G25)
Format: A4

A comprehensive guide to the installation and commissioning of direct and indirect fired appliances, radiant heating and gas equipment.

COMMERCIAL CATERING (G26)
Format: A4

Essential information on installing, commissioning and servicing commercial catering appliances.

To obtain further information and order any of the publications listed, contact Publications on: Tel: 01485 577800 / Fax: 01485 577758 / E-mail: publications@cskills.org / www.cskills.org/publications

LIQUEFIED PETROLEUM GAS (LPG)

LIQUEFIED PETROLEUM GAS SAFETY (G80) — Format: A4 in a ringbinder/A4

The industry reference manual for those working only on LPG systems. It covers all you need to know about ◆combustion ◆appliance gas safety devices and gas controls ◆principles of gas flues ◆flueing standards ◆ventilation requirements ◆emergency procedures ◆unsafe situations ◆warning notices and labels.

This pack consists of: ◆*Gas Safety (G1) pack*, ◆*Liquefied Petroleum Gas Safety (G18) book*, ◆*Liquefied Petroleum Gas Safety Course Workbook (G81)*.

LIQUEFIED PETROLEUM GAS SAFETY (G18) — Format: A4

The essential bolt-on to those working with natural gas and looking to extend into LPG. If you already own a *GAS SAFETY (G1)* pack, all you need is this book with its LPG-specific sections ◆installation ◆fire precautions and procedures ◆combustion ◆testing and commissioning installations ◆service pipework ◆bulk gas supply systems ◆the leisure industry.

ELECTRICAL

BS 7671: REQUIREMENTS FOR ELECTRICAL INSTALLATION (E1) — Format: A4 Wiro-bound

The standard reference book for electrical work. The easy-to-follow text, supported by diagrams, explains the complex regulations in terms a practical electrician can understand. It now incorporates reference to the IEE on-site guide that enables you to make calculations and design circuits in a much quicker and simpler manner.

ELECTRICAL INSTALLATION PACK (E3) — Format: A4 in a ringbinder

Over 430 pages of illustrated reference material divided into four sections:

- Basic Practical Skills – describes the tools required for electrical installation work and how to use them
- Wiring Installation Practice – deals with terminating cables, flexible cords and installing PVC cables, conduit trunking, MICC, SWA and FP200 wiring systems. (Complies with the 16th Edition *IEE Wiring Regulations*)
- Basic Electrical Circuits – covers standard circuit arrangements for lighting and power circuits, and relevant IEE Regulations
- Safety at Work – essential advice on safety at work, from securing ladders to dealing with electric shock. It also gives the key points of relevant Acts and Regulations.

ESSENTIAL ELECTRICS (E14) — Format: A4

An indispensable reference book for plumbers, gas fitters and heating and ventilating engineers whose work requires basic electrical knowledge and an understanding of electrical regulations.

CENTRAL HEATING CONTROLS (E15) — Format: A4

Deals with different types of central heating control systems for wiring and fault finding.

COMBINATION BOILERS (E19) — Format: A4

An invaluable reference manual for engineers who want to understand the principles of combination boilers. This manual covers most of the content for the ConstructionSkills Essential Electrics and Combination Boiler Fault Finding course. Over 80 pages of illustrated reference information covering ◆types of boilers ◆designs ◆wiring diagrams ◆installation ◆commissioning and servicing ◆fault finding.

WATER

UNVENTED HOT WATER STORAGE SYSTEMS (W2) — Format: A4

An informative guide for installing unvented hot water storage systems. It covers most of the content for the ConstructionSkills training and assessment scheme, including: ◆types of system ◆design ◆controls ◆installation ◆commissioning and decommissioning ◆servicing and fault diagnosis ◆relevant Building Regulations ◆good practice.

REFRIGERANTS

SAFE HANDLING OF REFRIGERANTS (R2) — Format: A4

Essential information, primarily designed for operatives undertaking ConstructionSkills Safe Handling of Refrigerants training and assessments, it covers ◆environmental impact ◆fluorocarbon control and alternatives ◆regulations ◆recovery and handling ◆refrigeration theory ◆good practice ◆automotive installations.

SAFE HANDLING OF ANHYDROUS AMMONIA (R4) — Format: A4

Essential information for handling anhydrous ammonia. Primarily designed for operatives undertaking ConstructionSkills Safe Handling of Anhydrous Ammonia training and assessments, it covers ◆safety and environmental issues ◆regulations ◆good practice.

PIPEWORK AND BRAZING (R6) — Format: A4

Primarily for operatives undertaking ConstructionSkills Pipework and Brazing training and assessments for refrigeration systems, it covers ◆health and safety ◆materials and equipment ◆lighting procedures.

To obtain further information and order any of the publications listed, contact Publications on: Tel: 01485 577800 / Fax: 01485 577758 / E-mail: publications@cskills.org / www.cskills.org/publications